COMPUTER-AIDED FORENSIC FACIAL COMPARISON

COMPUTER-AIDED FORENSIC FACIAL COMPARISON

edited by
Martin Paul Evison
Richard W. Vorder Bruegge

CRC Press
Taylor & Francis Group
Boca Raton London New York

CRC Press is an imprint of the
Taylor & Francis Group, an **informa** business

CRC Press
Taylor & Francis Group
6000 Broken Sound Parkway NW, Suite 300
Boca Raton, FL 33487-2742

CRC Press is an imprint of Taylor & Francis Group, an Informa business

Printed in the United States of America on acid-free paper
10 9 8 7 6 5 4 3 2 1

International Standard Book Number: 978-1-4398-1133-7 (Hardback)

Visit the Taylor & Francis Web site at
http://www.taylorandfrancis.com

and the CRC Press Web site at
http://www.crcpress.com

Table of Contents

Preface

The face is the medium by which we as human beings recognize and identify each other. Evidence from psychology suggests the ability to recognize faces relies on a specialized part of the brain, which has developed in response to an evolutionary pressure to reliably identify kith and kin, and distinguish them from strangers. Faces are universal and by their very nature remotely detectable. Experience leads us to believe they are unique. Faces appear to offer several of the ideal attributes of a method of human identification.

Facial identification is, however, a paradox. What seems to be an effortless process in the mind transpires to be an intractable one to emulate, especially for application in the courtroom. The jury requires a real and tangible means of assessing whether the face captured at the crime scene matches that of the defendant: something comparable to the Galton details of a fingerprint or the peaks plotted in an electropherogram generated in a comparison of DNA. Facial comparison is, however, uniquely complicated by the confounding influences of pose angle, lighting, lens distortion, focus, image resolution, image compression, and so forth. Even in the event of a postulated match—in which an assessment of error can also be provided—there remains the fundamental question of how many other individuals in the population would also match?

Inevitably, research—measurement, quantification of variation, model building and hypothesis testing—is the only conceivable route to solving these problems. In facial comparison, most of the technology and methods available in research are untried. The following represents an initial attempt to begin to address a number of the scientific and technical issues.

Martin Paul Evison and Richard W. Vorder Bruegge
June 30, 2009

The Editors

Martin Paul Evison, Ph.D., is an associate professor of anthropology at the University of Toronto, director of the University's Forensic Science Program, and a member of the Advisory Committee of the Centre for Forensic Science and Medicine. He has a Ph.D. in ancient DNA from the University of Sheffield and, from 1994 until 2005, he was the forensic anthropologist at Sheffield Medico-Legal Centre, where he led a research group focusing on computational and molecular methods of human identification. He has undertaken case work on five continents in forensic archaeology and anthropology, facial identification, and contamination or innocent transfer of DNA. He has been published in the *Journal of Forensic Sciences* and *Forensic Science Communications* among other journals, and is currently leading a research project examining genetic factors underlying normal face shape variation.

Richard W. Vorder Bruegge, Ph.D., has worked for the Federal Bureau of Investigation (FBI) since 1995. His work involves analyzing film, video, and digital images that relate to crime and intelligence matters, as well as testifying in court. He has conducted thousands of forensic examinations in hundreds of cases, and has testified as an expert witness in federal and state courts across the United States, as well as internationally. An intrinsic part of his job is to conduct research in the field of forensic image analysis, including facial identification. He was chair of the Scientific Working Group on Imaging Technology (SWGIT) from 2000 to 2006 and was elected chair of the Facial Identification Scientific Working Group (FISWG) in 2009. In 2007, he was invited to brief the National Academy of Sciences Committee on the Future of Forensic Sciences on the discipline of forensic photographic comparisons. His publications include a chapter in the *John Wiley Encyclopedia of Imaging Science and Technology* titled "Imaging Sciences in Forensics and Criminology," as well as articles published in forensics and biometrics literature. He has delivered close to 100 formal presentations at forensic science meetings and is a fellow of the American Academy of Forensic Sciences. The FBI has designated him as the bureau's point of contact for face and iris recognition.

Contributors

Alvaro Pallares Bejarano, Ph.D.
School of Physical Sciences
University of Kent
Canterbury, United Kingdom

Ian L. Dryden, Ph.D.
Department of Statistics
University of South Carolina
Columbia, South Carolina, USA

Martin Paul Evison, Ph.D.
Director, Forensic Science Program
Department of Anthropology
University of Toronto Mississauga
Mississauga, Ontario, Canada

Nick R.J. Fieller, Ph.D.
Department of Probability and Statistics
University of Sheffield
Sheffield, United Kingdom

Lorna Goodwin, Ph.D.
School of Psychology
University of Nottingham
Nottingham, United Kingdom

Edward Lester, Ph.D.
Department of Chemical and Environmental Engineering
University of Nottingham
Nottingham, United Kingdom

Xanthé D.G. Mallett, Ph.D.
Centre for Anatomy and Human Identification
College of Life Sciences
University of Dundee
Dundee, United Kingdom

Matthew I.S. Maylin, Ph.D.
School of Physical Sciences
University of Kent
Canterbury, United Kingdom

Lucy Morecroft
Department of Probability and Statistics
University of Sheffield
Sheffield, United Kingdom

Damian Schofield, Ph.D.
Department of Computer Science
SUNY Oswego
Oswego, New York, USA

Christopher J. Solomon, Ph.D.
School of Physical Sciences
University of Kent
Canterbury, United Kingdom

Richard W. Vorder Bruegge, Ph.D.
Digital Evidence Laboratory
Quantico, Virginia, USA

Acknowledgments

Provisionally coined "IDENT—An anthropometric method of facial comparison," this research program was originally conceived of as a research and development project supported by Sheffield University Enterprises Limited (SUEL), UK. A research proposal (IS-R-946-SUFP-FP) was submitted to the Technical Support Working Group (TSWG) in response to solicitation DAAD05-02-Q-4738, operational requirement number 946: "Photogrammetric Matching of Faces." Martin Paul Evison would like to express thanks to SUEL, and especially to business mentor John Evans, for their support during the development of the proposal, and also to the South Yorkshire Police and the UK Police Information Technology Organisation (now part of the National Policing Improvement Agency) for valuable feedback and encouragement. The research program was funded by TSWG (Project T216E) according to a requirement of the Federal Bureau of Investigation Forensic Audio, Video, and Image Analysis Unit. We would like to thank Jeff Huber and colleagues for their support and guidance.

Data collection commenced in December 2003 at the Magna Science Adventure Centre, Rotherham, UK. We owe a special thanks to the Magna staff, the research technicians who assisted with data collection, and over 3000 members of the public who generously volunteered to let their 3D facial images be used in research for crime prevention and detection.

In addition to the collaborating coauthors of this volume, we would like to thank numerous colleagues and former graduate students from the University of Sheffield School of Medicine and Biomedical Sciences for their contributions and, especially, the staff of Sheffield Medico-Legal Centre.

Finally, we thank Dr. Gary Dickson, who was responsible for the applications programming work, and for undertaking routine photography. Gary can be credited with the development of the software prototypes used in research and testing, as well as contributing substantially to the overall collaborative effort.

Introduction

1

MARTIN PAUL EVISON AND
RICHARD W. VORDER BRUEGGE

Contents

The faces of crime and terrorism are commonplace. Unlike DNA and fingerprints, which are normally encountered only at crime scenes or in government databases, images that depict the faces of criminals, terrorists, and their victims are ubiquitous. The ubiquity of digital cameras, cell phone cameras, surveillance video systems, and webcams, to say nothing of traditional film and broadcast video, results in the generation of countless facial images every day. Furthermore, the ease with which facial images can be transmitted, shared, and stored through the Internet and on devices such as servers, digital video recorders, computer hard drives, and removable media makes it possible to access millions of images and videos depicting faces with just a few clicks of a button.

As a result, law enforcement and intelligence agencies have many more opportunities to acquire and analyze images that depict persons of interest, whether they may be suspects of a crime, witnesses, or victims. In most cases, such images are used for investigative or recognition purposes, wherein an investigator or witness will look at a photograph and because of a prior association or familiarity with the subject, "recognize" the individual and thus be able to "identify" them. In some cases, however, the identity of the individual depicted in an image is subject to debate. In these cases, analysis by an expert may be necessary to either confirm or exclude a specific individual as being the subject depicted in an image.

1.1 Forensic Facial Comparison

1.1.1 Background

A reliable method of confirming or excluding facial identity from photographs—"forensic facial comparison"—has an important role in criminal and intelligence investigations. Işcan (1993) discusses two major categories for photographic facial comparison: "morphological" and "anthropometric." A third category identified by Işcan as "video superimposition" should be considered as a combination of the first two—an examiner resizes and overlays two facial images, then determines whether there is an alignment and correspondence of features. Although explicit measurements may not be recorded, any differences in the geometric arrangement of features after the alignment is achieved must be incorporated into the ultimate conclusion. This method appears to be the preferred technique of forensic practitioners today.

Morphological analyses involve the direct comparison of facial features, and require that the expert/examiner identify similarities and differences in the observed characteristics. These characteristics include both "class" and "individual" characteristics, where the former can represent characteristics common to many individuals (e.g., the overall shape of the nose, eyes, or mouth), while the latter can include such characteristics as freckles, moles, and scars. The location and distribution of these characteristics are considered in morphological analyses, but are not explicitly measured. This type of analysis is discussed in Spaun and Vorder Bruegge (2008).

Anthropometric analyses rely upon the explicit measurement of landmarks on the face and a comparison of these measurements between the questioned and known subjects. In most cases, blemishes such as freckles, moles, or scars are not incorporated into anthropometric analyses.

While a combination of the morphological and anthropometric techniques through overlay analysis is now anticipated to offer the best possible mechanism for photographic identification of individuals by human examiners, the anthropometric technique alone is the focus of the research described herein. Examples of the use of anthropometric analysis in criminal cases have been described by Halberstein (2001).

Any anthropometric analysis of images requires the incorporation of photogrammetric principles. This stems from the basic principle that while most images are two-dimensional (2D) representations, the face is a three-dimensional (3D) object. Hence, the research summarized in this volume was undertaken with the ultimate aim of creating, a reliable method of photogrammetric analysis of faces that can be used to confirm or exclude an identity based on facial measurements—craniofacial anthropometrics—as an operational capability for both criminal investigation and prosecution.

At a more basic level, however, a fundamental question addressed by this research is whether human beings may be distinguished from one another based solely upon the geometrical arrangement of facial landmarks common to all. Whereas it is accepted that all human beings may be distinguished from one another on the basis of their DNA (with the exception of identical twins or multiples), and probably through other modalities of human identification such as fingerprint and iris patterns, the ability to discriminate every human being from another based upon their facial characteristics is an open question—and one that needs to be answered. The recent National Academy of Sciences report, *Strengthening Forensic Science in the United States* (NAS 2009) did not explicitly address forensic facial comparison, but could have been when it stated "[i]n most forensic science disciplines, no studies have been conducted of large populations to establish the uniqueness of marks or features. ... A statistical framework ... is greatly needed."

To date, no statistical means of individual identification from the face has been developed for use in court. Current photogrammetric methods intended to offer proof of facial identification are crude and may lead to erroneous matches or exclusions; such errors have the potential to result in both unwarranted convictions and acquittals of guilty suspects. It with this in mind that the investigation described in this chapter was undertaken.

1.1.2 Overview of Investigation

The investigation can be conceptually divided into two parts: (1) collection of a database of 3D faces and face measurements and (2) provision of a rapid and automatic means of recording, comparison, matching, and exclusion of facial images that is defensible in a court of law on the basis of empirical scientific evidence open to peer review and scientific scrutiny. In practice, however, this volume is organized into chapters describing self-contained but interrelated investigations and database and prototype development exercises.

The evaluation of instruments for 3D face capture is described in Chapter 2. Investigations of anthropometric face shape variation in 3D and collection of a large database sample of 3D faces are described in Chapters 3 and 4, respectively. Landmark variation in 2D is investigated in Chapter 5. Each chapter offers consideration of error and repeatability in measurement. Wider issues of landmark visibility in 3D, and error due to the influence of lens and perspective are considered independently in Chapters 6 and 7. Chapters 8 and 9 explore more theoretically orientated avenues of related research, specifically landmark placement using the active shape model and estimation of missing landmark position using the EM algorithm. Chapter 10 introduces the fundamental issue of courtroom admissibility. Chapter 11 describes the prototypic application toolset developed as a consequence of the preceding research and the results of preliminary testing. Finally, Chapter 12 reviews the overall problems and prospects of the effort.

Both commercial and custom software application tools were used in the project. The personal computer platforms employed were Microsoft® Windows-based IBM-compatible PCs capable of hosting distributed low volume graphics-orientated systems for research, systems development, and testing purposes. Standard platforms consisted of 3.0 GHz Pentium 4 CPUs with 2 GB of RAM and 256 MB PCI-E dual-head graphics cards supporting two 19" LCD monitors at 1080 × 1045 pixel resolution. Statistical programming was undertaken in R (R 2009), and Microsoft Windows applications were developed in C++ and Visual Basic.

1.2 Ethical Approval and Informed Consent

Collection of information of any kind from human subjects for research purposes is regarded as highly ethically sensitive, and the collection of identifiable personal information—of which the face is clearly an emotive example—particularly so. As well as being of general importance, there was a specific requirement for data collection with informed consent and for steps to be taken to safeguard and preserve the strict anonymity of the public volunteers' identities. The facial images remain unobscured, however, to facilitate access by other bona fide researchers.

The research project and proposed data collection process was reviewed by a University of Sheffield research ethics committee. The review addresses issues such as scientific quality, potential for harm to volunteers or the public, informed consent, appropriateness of the information collected to the research goals, and confidentiality. Mechanisms for the control of access to the database, now and in the future, were regarded as critical issues.

Informed consent hinged on the appropriateness of information offered to volunteers, and details of data requested and consent given. Favorable ethical review was given, based on the information sheet, biographic information form, and consent form used in the data collection process. These are included in Appendix A.

It should be noted, in particular, that public volunteers' faces are to remain confidential and are not to be published or made public in any way. Volunteers may withdraw from the database at any time without giving a reason, and their database records will be deleted. Volunteers' data may be used in the future by bona fide researchers in crime prevention and detection—subjectively defined by forensic scientists as researchers in public organizations with a genuine interest in forensic facial identification.

Volunteers' ages were restricted to over 14. For 16 to 18 year olds, informed consent of the parent or guardian was requested and, if possible, obtained.

Departure from the information given to volunteers and uses consented to by them may constitute a breach of human subject research ethics.

1.3 Data Collection Program

The database and data collection program (Chapter 4) were designed by and undertaken under the direction of the Principal Investigator, who identified the Magna Science Adventure Centre (Rotherham, South Yorkshire, UK) as a promising site for data collection from public volunteers. Magna is a popular science museum located in a former steel foundry. To mitigate risk, however, plans were made for data collection at other venues and a partly mobile data collection platform was implemented based on the Geometrix FaceVision® FV802 Series Biometric Camera (see Chapters 2, 3, and 4). Similarly, a Cyberware® 3030PS Head and Neck Scanner—which uses an alternative laser-based technology—was installed in reserve (see Chapter 2).

In order to foster public interest in the research, a launch was held at the Magna Science Adventure Centre, with the invaluable support of representatives of the Federal Bureau of Investigation Forensic Audio, Video and Image Analysis Unit (FAVIAU; see Figure 1.1). The launch was a very successful

Figure 1.1 FBI visitors at the Magna Science Adventure Centre, Rotherham, UK.

event, attracting all the main network radio and television news services in the United Kingdom, and transpired to be more popular even than the opening of the Magna Science Adventure Centre itself. Although there was some interest in the privacy aspects of the research the popular interest and press coverage was overwhelmingly positive, and this trend continued throughout the project.

Potential volunteers were provided with the project information sheet and given the opportunity to discuss the project with technicians employed to assist with data collection. Members of the public willing to volunteer were asked to complete the biographic information sheet and consent form, which optionally allowed volunteers to give an email address via which they could be contacted with further information about the project. Volunteers were then scanned with the 3D stereophotographic and laser scanners.

Both scanners were commissioned, calibrated where necessary, and used in accordance with the manufacturers' instructions. Technicians undertook a formal induction and training process to ensure the manufacturer's procedures were followed, supported by site visits from representatives of both companies.

The Geometrix FaceVision FV802 Series Biometric Camera is portable, enabling data collection to be extended beyond the Magna site. These excursions had the added benefit of making the research project and methods employed known to other experts, fostering greater acceptance of the research within the forensic science community and permitting field testing of the 3D scanning equipment.

The data collection program was designed to meet a specific requirement for 3D data collection from a small representative sample population stratified by sex, age, build, and geographic ancestry. The biographic information sheet was used to record the individual's sex, age, and ancestry—self-assessed subjectively according to the U.K. National Census classifications. The site location, scanner used, and technician undertaking the scan were also recorded. In this way, analysis of craniofacial variation could be undertaken according to biological (sex, age) or bio-cultural (ancestry and intra- and inter-observer) site, observer, and scanner categories.

To ensure that the database size requirement was met, an excess of volunteers' faces were scanned. After some records were deleted because of quality control or other issues, over 3000 volunteers' facial images were obtained, the overwhelming majority collected with the Geometrix FaceVision FV802 Series Biometric Camera.

Each volunteer was provided with a copy of their own 3D facial image model generated with the Geometrix FaceVision software (Figure 1.2). The volunteer was then able to view their facial image in a Web browser on their own desktop using the Viewpoint Media Player (www.viewpoint.com).

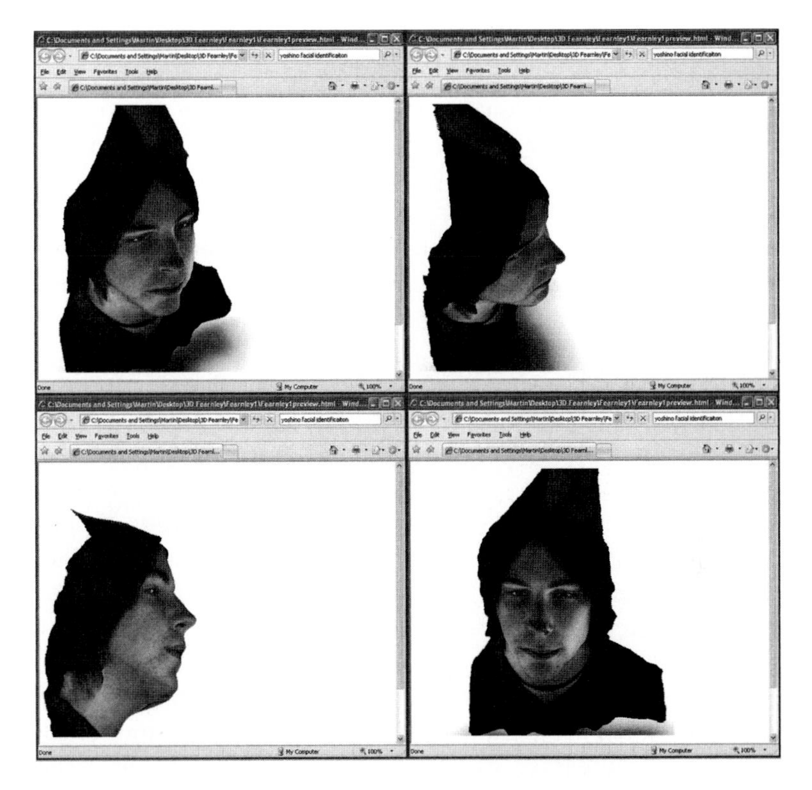

Figure 1.2 Screen shots of the 3D surface of a face captured using the Geometrix FaceVision FV802 Series Biometric Camera and viewed in Internet Explorer® with the Viewpoint Media Player.

1.4 Dissemination

1.4.1 Large Database Sample

The large sample database was retained in the following form:

- Geometrix ForensicAnalyzer® project folders, each containing the eight facial images of a volunteer in JPEG, MTS, and DXF format files of the 3D surface of the volunteer's face generated by the Geometrix FaceVision application tool, and a calibration file permitting regeneration of the face surface using FaceVision. There are over 3000 project folders with names corresponding to the unique key references of the volunteers scanned with the Geometrix FaceVision FV802 Series Biometric Camera. Each folder contains 12 files: eight JPEG files corresponding to the eight 2D digital camera images, a calibration file, a landmark data file, and the MTS and DXF format 3D face surface image data.

- Cyberware project folders each containing the facial image of a volunteer in DXF format. The folder name corresponds to the unique key reference of the volunteers scanned using the Cyberware equipment. The folder contains DXF and RGB 3D surface and color image files, respectively.

The large sample database is available to bona fide researchers in crime prevention and detection via intergovernmental agreement (see Chapter 4).

1.4.2 Digital Media: Landmark Datasets and Prototypes

The following landmark datasets, research material, and prototypic programs are included on the digital media included with this volume:

- CSV format file containing all landmark data collected from the 3115 volunteers landmarked using Geometrix FaceVision, including those with "missing data"

  EVB_real_data.csv

- CSV format file containing landmark data collected from the 3115 volunteers landmarked using Geometrix FaceVision, but excludes those with "missing data"

  EVB_complete_real_data.csv

- CSV format file containing landmark data collected from the 3115 volunteers landmarked using Geometrix FaceVision, where values are generated for "missing data" using the EM algorithm

  EVB_em_data.csv

- CSV format file containing the biographic data for all volunteers

  EVB_biographic_data.csv

- CSV format file containing a list of the 30 landmarks and their names

  EVB_final_30_landmarks.csv

- Folders containing copies of Microsoft Windows programs, R programs, and MATLAB® and MAXscript® programs, corresponding to application tools and research tools used in each chapter

1.4.3 Technology Transition

The investigation has resulted in prototypic models warranting further development, with the ultimate aim of technology transition from research to application. Commercial use of such research—including commercial and open source software—may require special licensing arrangements to be made with the supplier.

References

Halberstein, R. A., 2001. The application of anthropometric indices in forensic photography: Three case studies. *J. Forensic Sci.* (46)6: 143841.

Işcan, M. Y., 1993. Introduction to techniques for photographic comparison: Potential and problems. In *Forensic analysis of the skull*, ed. M. Y. Işcan and R. P. Helmer. New York: Wiley-Liss, 258 pp.

NAS, 2009. *Strengthening forensic science in the United States: A path forward.* Washington, DC: National Research Council, 254 pp.

R., 2008. *The R project for statistical computing.* http://www.r-project.org (accessed June 30, 2009).

Spaun, N. A., and R. W. Vorder Bruegge, 2008. Forensic Identification of People from Images and Video. *Proceedings of BTAS 2008,* 13.

Image Quality and Accuracy in Three 3D Scanners

2

LORNA GOODWIN, MARTIN PAUL EVISON, AND DAMIAN SCHOFIELD

Contents

2.1 Introduction

There are a number of 3D stereophotographic and laser-based scanners available commercially. Of these, this project assessed three for 3D image quality and accuracy. These were a Cyberware® 3030PS Head and Neck Scanner (Cyberware, Inc., Monterey, CA), a Geometrix FaceVision® FV802 Series Biometric Camera (ALIVE Tech, Cumming, GA), and a 3dMDface™ System (3dMD, Atlanta, GA) scanner.

These provisional assessments were intended to provide guidance as to the relative benefits of the equipment for the purposes of computer-assisted forensic facial comparison and identification factors affecting their optimal use.

All three scanners were investigated via a visual assessment of 3D image quality and an analysis of accuracy of anthropometric landmarking from these images. Assessment of the 3dMDface System scanner, however, was restricted to the three subject images that were available. These subjects were also scanned with the Cyberware 3030PS Head and Neck Scanner and the Geometrix FaceVision FV802 Series Biometric Camera, permitting a limited direct comparison between all three scanners. As the assessment of the 3dMDface System scanner is based on a very small sample, it should be treated with circumspection.

Data was collected from the three scanners and directly from live subjects using calipers for the purposes of comparative controls.

It is important to note that neither the Cyberware 3030PS Head and Neck Scanner or the 3dMDface System is built for forensic facial comparison. The Geometrix FaceVision FV802 System, however, offers a further separate and purpose-designed means of 3D landmarking via triangulation from series of 2D images. This capability is not evaluated in this chapter, which is concerned with 3D image quality and accuracy; it is considered in detail in Chapter 3, however.

2.2 Cyberware 3030PS Head and Neck Scanner

The Cyberware 3030PS Head and Neck Scanner is shown in Figure 2.1.

The scanner uses a low intensity laser light source, which is projected at the subject's head on a rotating platform. As the platform rotates, the range and color of the lighted points of the head are captured by sensors, and are streamed into two data files: one containing the 3D coordinates of the head and the second a color or texture map.

2.2.1 Method of Assessment

Performance of the Cyberware 3030PS Head and Neck Scanner was evaluated in two ways: first, via a subjective assessment of 3D image quality and, second, via an empirical comparison of measurements collected from 3D scanned images with those collected directly from a live subject and two artificial head-shaped objects (mannequins) using calipers.

Thirty-two 3D image datasets of the same subject were captured at the high resolution (512 × 512 coordinate matrix) setting of the scanner, following the manufacturer's instructions. Two mannequin heads, one polyvinylchloride (PVC) and one polystyrene, were scanned 15 times each in the same way in order to provide comparative controls.

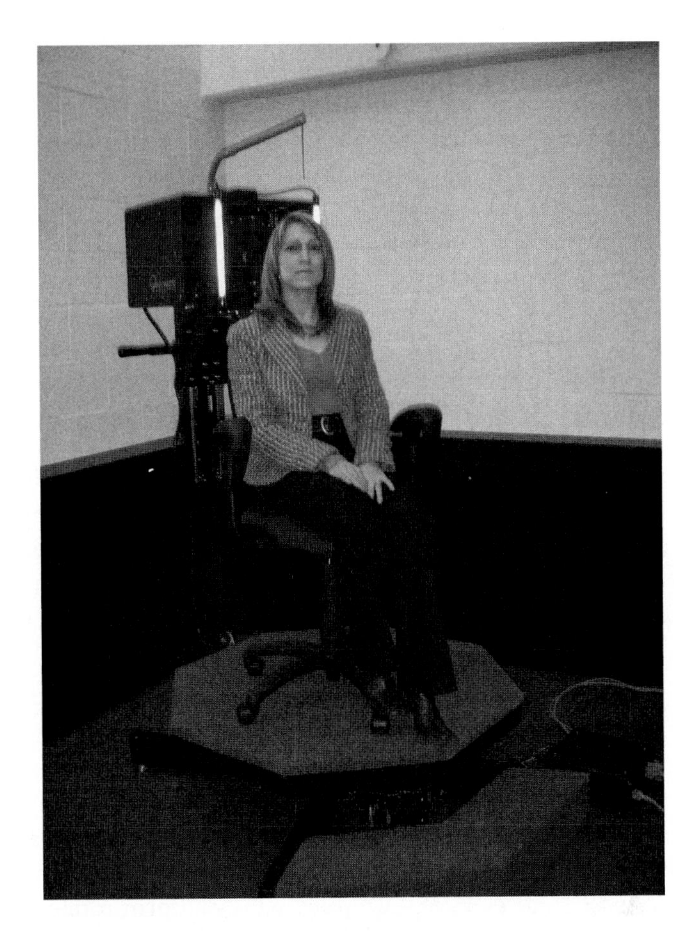

Figure 2.1 Cyberware 3030PS Head and Neck Scanner.

Each dataset was then exported to the 3D modeling package 3ds Max® version 7.0 (Autodesk®, San Rafael, CA) in 3ds format, and seven anthropometric landmarks were placed on each wireframe model: the pronasale (prn), sellion (se), pogonion (pg), subnasale (sn), sublabiale (sl), and exocanthion left (ex l) and right (ex r) landmarks (Figure 2.2).

The Euclidean distances for every combination of two landmarks (21 pairs in total) were measured for each of the 32 datasets representing the live subject, and each of the 15 datasets representing each of the two mannequin heads.

Three repetition measurements of these 21 pairwise distances were collected directly from the two mannequin heads using digital calipers (Mitutoyo USA, Aurora, IL) by the same observer and from the live subjects and also by a separate observer.

Figure 2.2 The seven landmarks used in the assessment. (Bilateral landmarks are indicated by an asterisk suffix.)

Image quality was assessed by careful visual scrutiny of each texture map and 3D wireframe surface. In order to assess accuracy of the 3D geometry, pairwise distance measurements collected from the scanned datasets were analyzed and compared with the measurements collected with calipers.

2.2.2 Assessment of Image Quality

A subjective assessment of 3D image quality revealed a series of imperfections in some of the 3D images. Stretching (Figure 2.3) is a consequence of the obtuse angle of incidence of the laser during scanning. Spikes (also Figure 2.3) appear to be the result of specular highlighting of the skin or reflection of the laser by the moist sclera of the eye. Scan overlap results from slight movement of the subject as they are being scanned (see Figure 2.4).

Stretching was also observed on 3D images collected from the PVC and polystyrene mannequin heads (Figures 2.5 and 2.6, and 2.9, respectively).

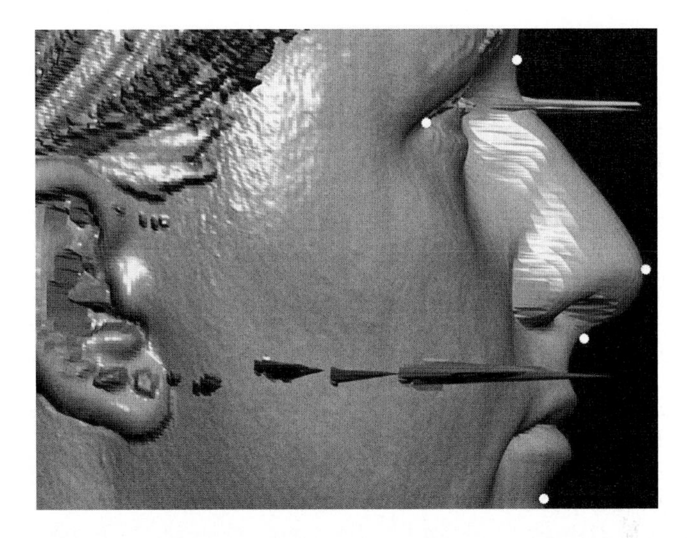

Figure 2.3 Nose stretching and spikes are visible.

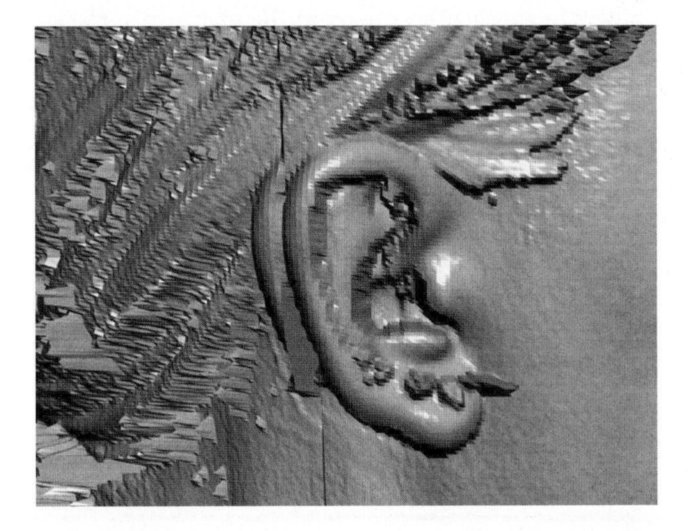

Figure 2.4 Scan overlap at the ear.

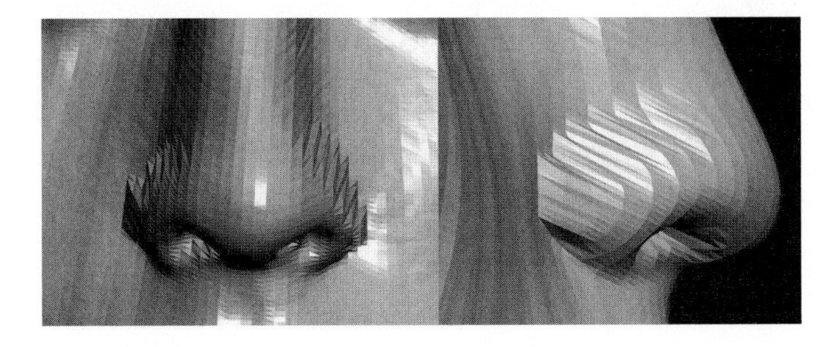

Figure 2.5 Nose stretching on the image of the PVC mannequin.

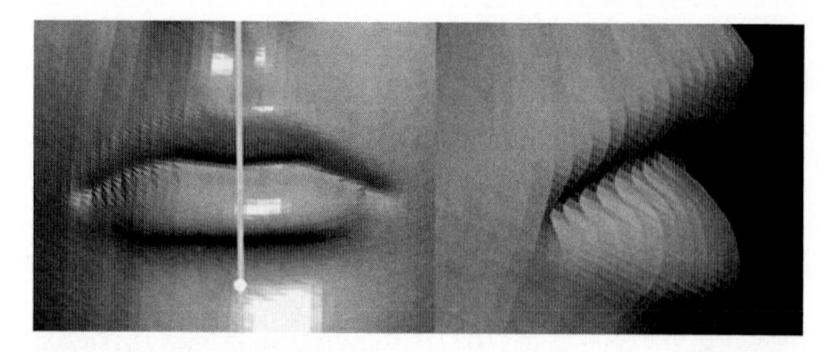

Figure 2.6 Lip stretching on the image of the PVC mannequin (left: note the distortion relative to the facial midline).

Other anomalies detected on the scanned PVC image included spikes (Figure 2.7). Ridges, which were not bilateral, were detected on both the scanned PVC (Figure 2.8) and polystyrene (Figure 2.10) images. These are assumed to be due to a lighting bias or the particular sensitivity of a lightweight object to subtle unevenness in the platform surface. The manufacturer's instructions emphasize the need for the platform to be level.

Spikes were not observed on the scanned polystyrene image, probably as a consequence of its relatively lusterless surface. In contrast, nose stretching was more striking—an observation attributed to the stepped topology of the polystyrene mannequin head.

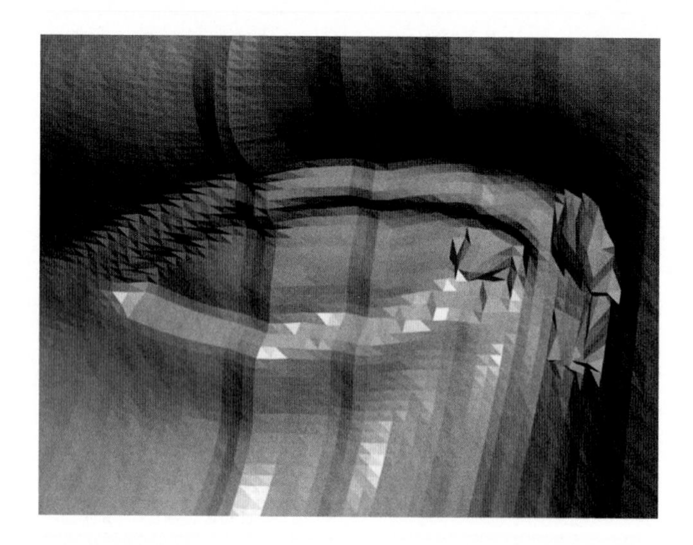

Figure 2.7 Spikes on the image of the PVC mannequin (also note ridges).

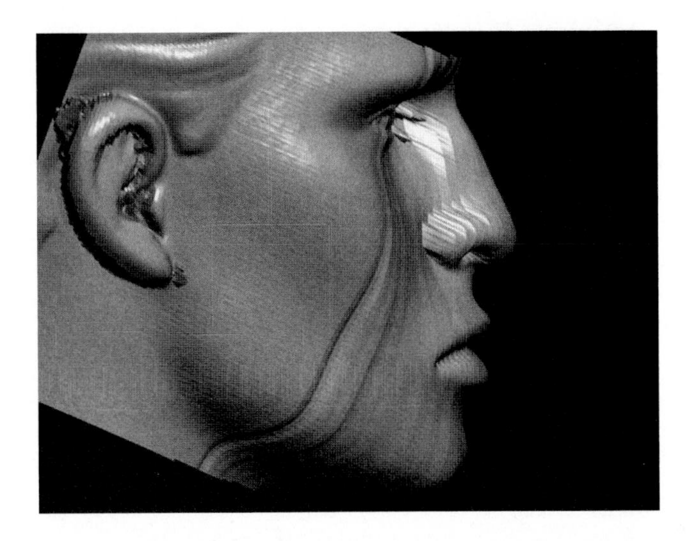

Figure 2.8 Ridges on the image of the PVC mannequin (also note spikes at the ear).

2.2.3 Comparison of Pairwise Measurements

A chart illustrating variation in pairwise distance between landmark measurements collected from scanned images compared with average distances collected with calipers from live subjects is shown in Figure 2.11.

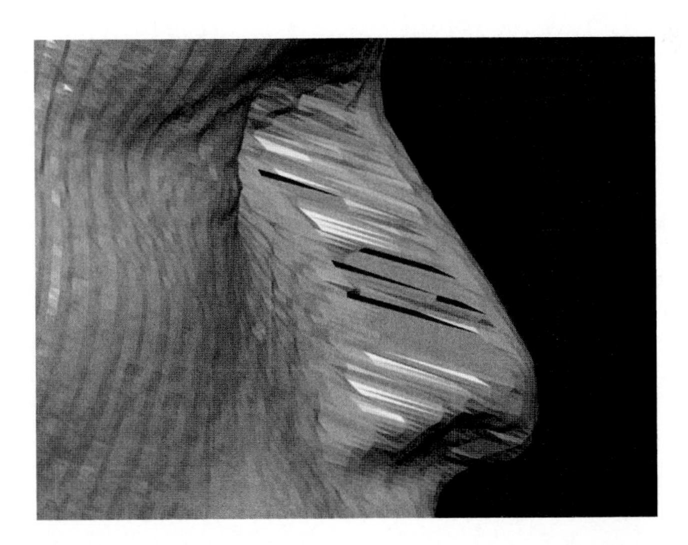

Figure 2.9 Nose stretching on the image of the polystyrene mannequin.

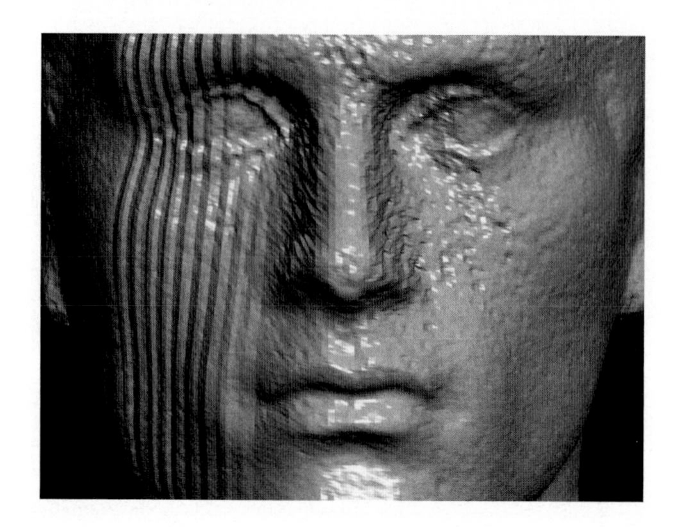

Figure 2.10 Ridges on the image of the polystyrene mannequin.

A chart illustrating variation in pairwise distance measurements collected from scanned images compared with average distances collected with calipers from the PVC mannequin is shown in Figure 2.12.

A chart illustrating variation in pairwise distance measurements collected from scanned images compared with average distances collected with calipers from the polystyrene mannequin is shown in Figure 2.13.

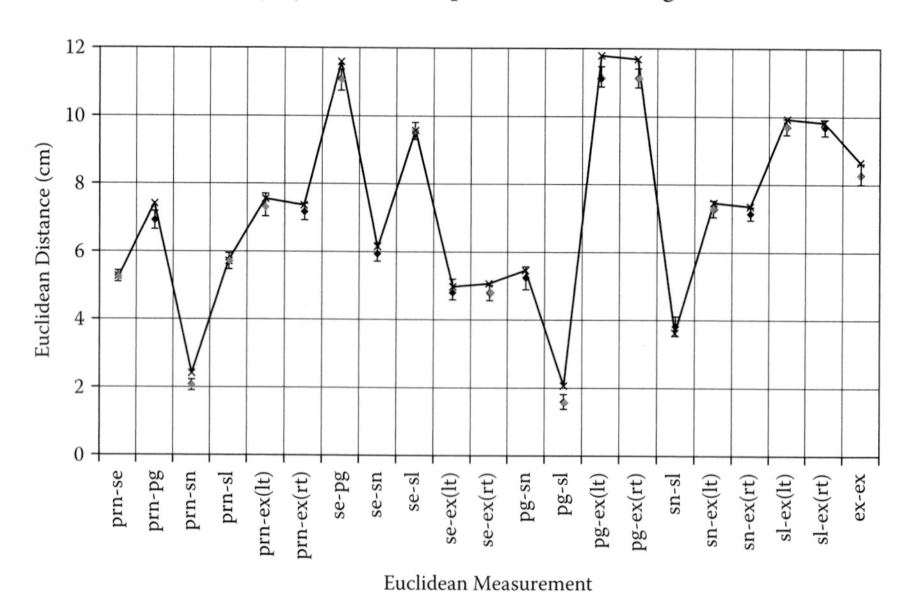

Figure 2.11 Chart showing minimum, maximum, and average pairwise distances between landmarks collected from scanned images (bars) compared with averages collected with calipers from the living subject (line).

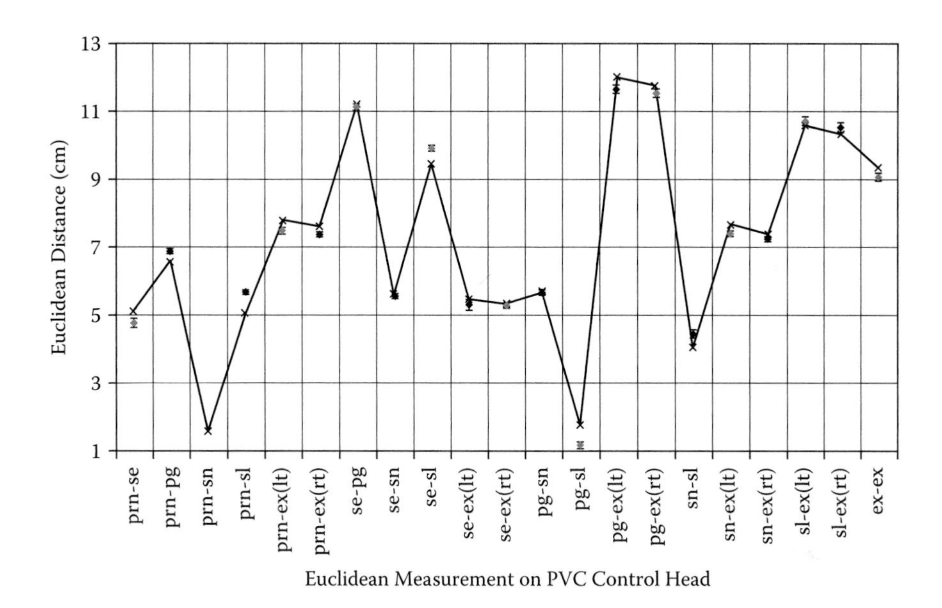

Figure 2.12 Chart showing minimum, maximum, and average pairwise distances between landmarks collected from scanned images (bars) compared with averages collected with calipers from the PVC mannequin (line).

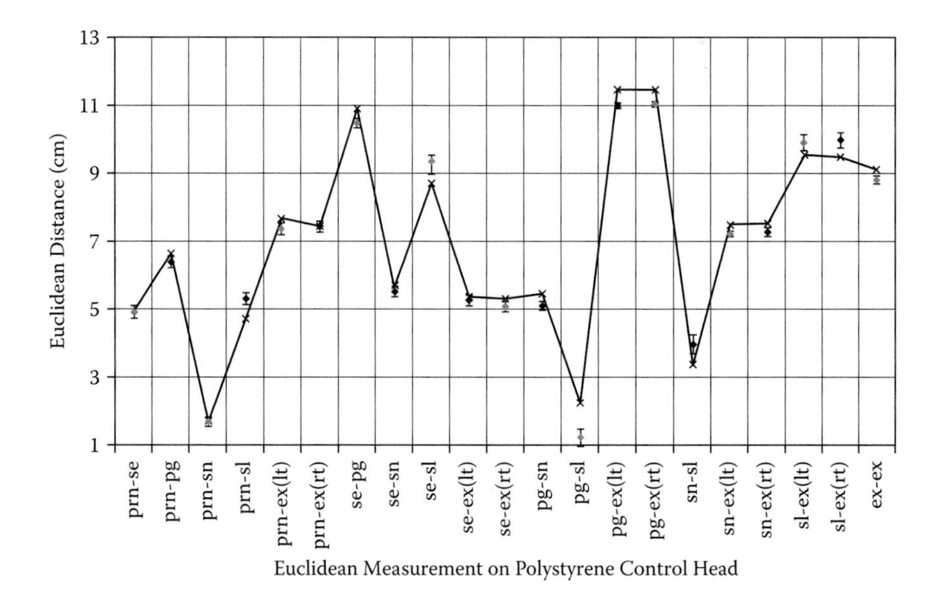

Figure 2.13 Chart showing minimum, maximum, and average pairwise distances between landmarks collected from scanned images (bars) compared with averages collected with calipers from the polystyrene mannequin (line).

2.2.4 Discussion of Cyberware Scanner Performance

Anomalies in the scanned surface will affect the position of any anthropometric landmarks derived from it. Bush and Antonyshyn (1996) showed that variation in head position and inclination had a significant effect on the 3D image due to alteration in the orientation of the curved surfaces of the face with respect to the plane of light projected by the laser scanner. The anticipated effect of head inclination on the reliability of landmark positioning was specific for each landmark. Spikes and stretching or ridges are each likely to affect landmark position, and certain landmarks will be affected more than others. Any movement of the subject may affect the quality of the scanned surface.

The largest variation detected in measurements from scans of live subjects was about 7 mm for the pogonion to subnasale (pg-sn) distance. In a study on the repeatability of landmarking from laser-scanned faces, Coward et al. (1997) reported comparable coefficients of repeatability with error measurements ranging between 1.6 and 7.0 mm. Comparison between measurements derived from scans and those collected from live subjects indicates a small amount of error, which barely exceeds the range of variation in measurements of scanned images (Figure 2.11). The differences (in the region of a few millimeters) are small, and may be attributed to random error, inter-observer error in landmark positioning, or systematic error resulting from the two strikingly different modes of data collection.

Variation in pairwise measurement from scanned PVC mannequin appears to be less than that observed for live subjects, and differences from the averages collected from the real object tends to be less (Figure 2.12). Measurements collected from the polystyrene mannequin (Figure 2.13) shows more variation and differences, however. Since all of the measurements collected from the mannequins were collected by the same observer, little inter-observer error in landmark positioning is anticipated—although one landmark, the sublabiale (sl), appears to be implicated as observer error in the most divergent pairwise measurements. In any event, the findings appear to reveal systematic differences due to the surface properties of the materials from which the mannequins are constructed.

2.3 Geometrix FaceVision FV802 Biometric Camera

The Geometrix FaceVision FV802 Biometric Camera is shown in Figure 2.14.

The scanner uses eight one-megapixel 2D digital cameras held in fixed orientation on a gantry in order to capture eight 2D image files that cover the facial surface of the subject. The Geometrix FaceVision software uses areas of resemblance between groups of pixels in each of these images to synthesize a wireframe dataset representing the 3D facial surface. A setting in the software can be selected that allows the resolution to be determined in relation to the size of the wireframe mesh in polygons.

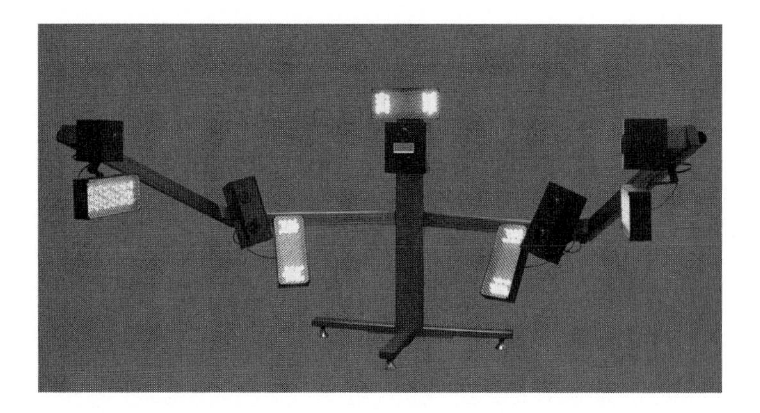

Figure 2.14 Geometrix FaceVision FV802 Series Biometric Camera.

2.3.1 Method of Assessment

Performance of the Geometrix FaceVision FV802 Series Biometric Camera was evaluated in two ways: first, via a subjective assessment of 3D image quality and, second, via an internal comparison of measurements collected from 3D scanned images captured using different sized polygon meshes.

Eight 3D image files of the same subject were captured at the four different resolution settings of the scanner in the 3ds format following the manufacturer's instructions. The resolutions were selected using the polygon number settings of the scanner; 20,000, 40,000, 60,000, and 80,000 polygon settings were chosen.

Each of the resulting 32 3ds files was then imported into 3ds Max and seven anthropometric landmarks were placed on each wireframe model, namely the pronasale (prn), sellion (se), pogonion (pg), subnasale (sn), sublabiale (sl), and exocanthion left (ex l) and right (ex r) landmarks (see Figure 2.2).

The Euclidean distances for every combination of two landmarks—21 pairwise distances in total—were measured for each of the 32 files.

In order to assess accuracy of 3D geometry, pairwise distance measurements collected from the scanned files were analyzed, and the findings at different polygon number settings was compared.

2.3.2 Assessment of Image Quality

A subjective assessment of 3D image quality revealed a series of anomalies. Texture map misalignment with the 3D wireframe (Figures 2.15 and 2.16) appears to be a reoccurring issue affecting the orbital region at every scan resolution used, but varying in extent with each scan. Other, apparently systematic, distortion affects the nose (Figure 2.17), and random distortion affects other parts of the facial surface (Figure 2.18).

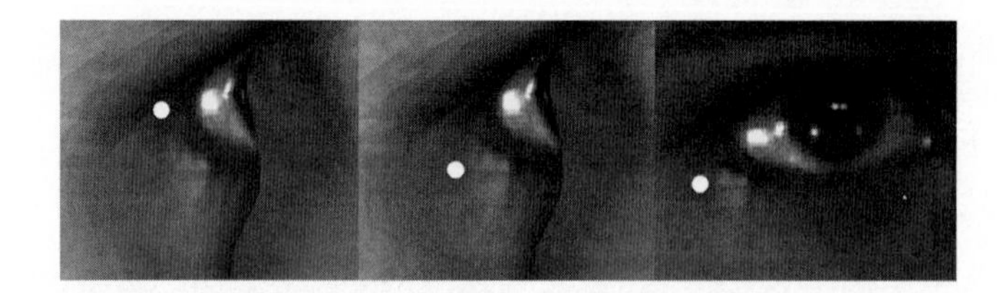

Figure 2.15 Illustration of discrepancy between 3D wireframe surface and texture map at the right exocanthion. Left: position of exocanthion on texture map. Center and right: position of exocanthion on 3D wireframe surface—scan at 60,000 polygon resolution. The difference between the wireframe and texture map position of the exocanthion is 4 mm. This value apparently varies with each scan.

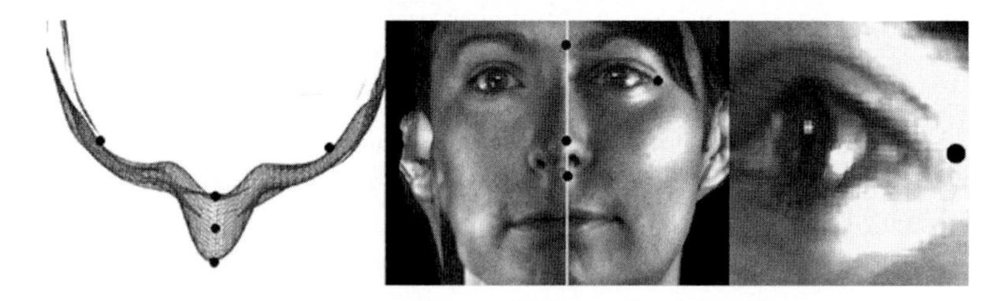

Figure 2.16 Illustration of discrepancy between 3D wireframe surface and texture map along the facial midline and at the left exocanthion. Left: position of landmarks on wireframe. Center: imposition of landmark geometry on texture map illustrating misalignment. Right: position of exocanthion on 3D wireframe surface showing slight misalignment (see also Figure 2.15).

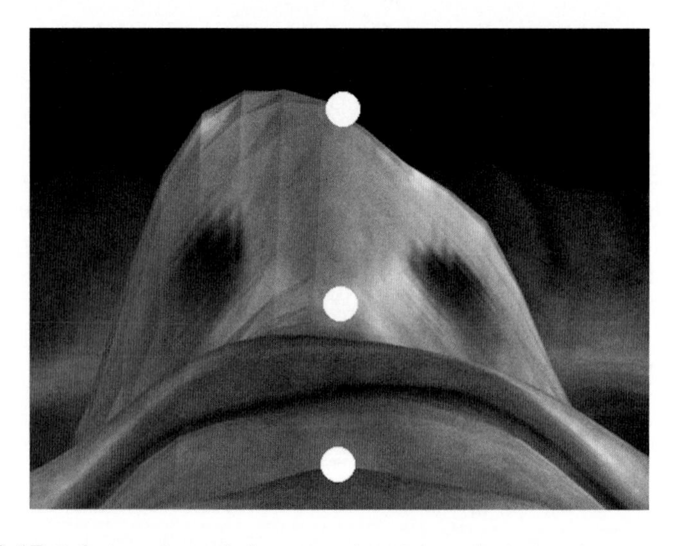

Figure 2.17 Inferior view of the nose showing a deviation in position of the nasal tip from the facial midline—scan at 80,000 polygon resolution.

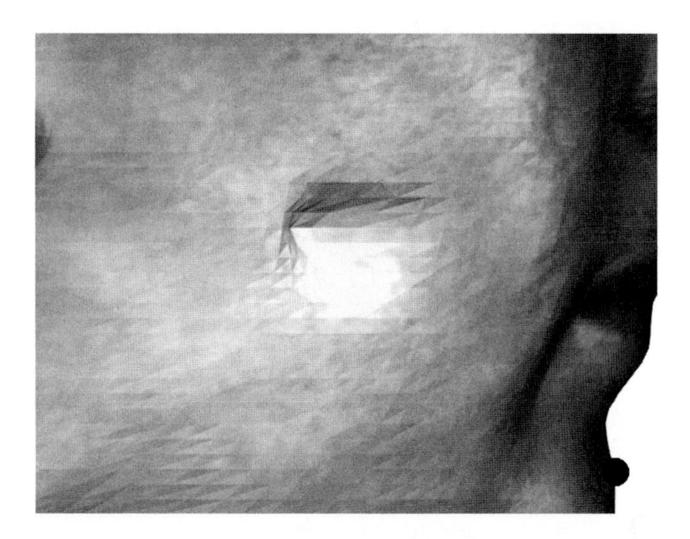

Figure 2.18 Right lateral view of the face showing a localized anomaly in the 3D surface—scan at 80,000 polygon resolution.

2.3.3 Comparison of Pairwise Measurements

Charts illustrating variation in pairwise distance measurements collected from scanned images at 20,000, 40,000, 60,000, and 80,000 polygons are shown in Figures 2.19 to 2.22.

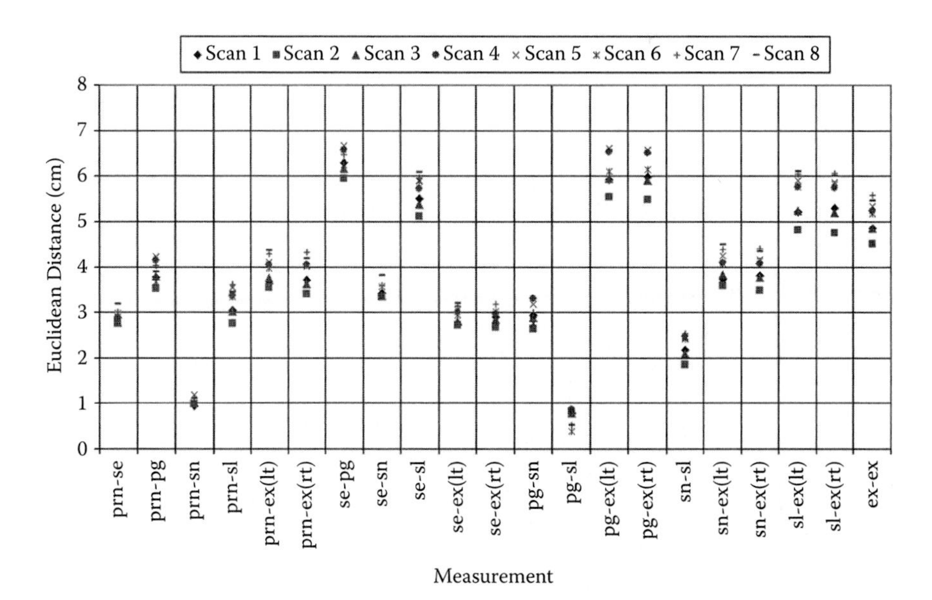

Figure 2.19 Chart showing minimum, maximum, and average pairwise distances between landmarks collected from scanned images at the 20,000 polygon resolution setting.

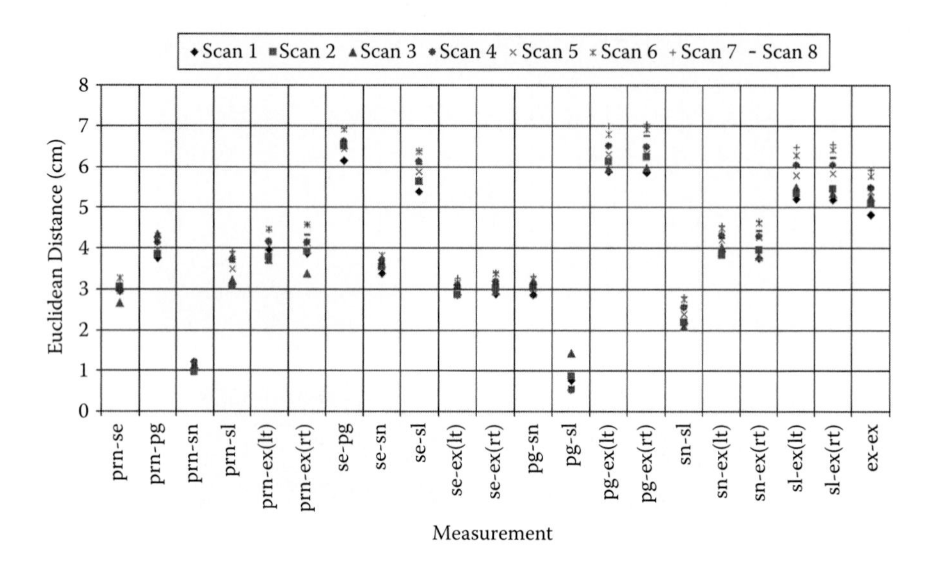

Figure 2.20 Chart showing minimum, maximum, and average pairwise distances between landmarks collected from scanned images at the 40,000 polygon resolution setting.

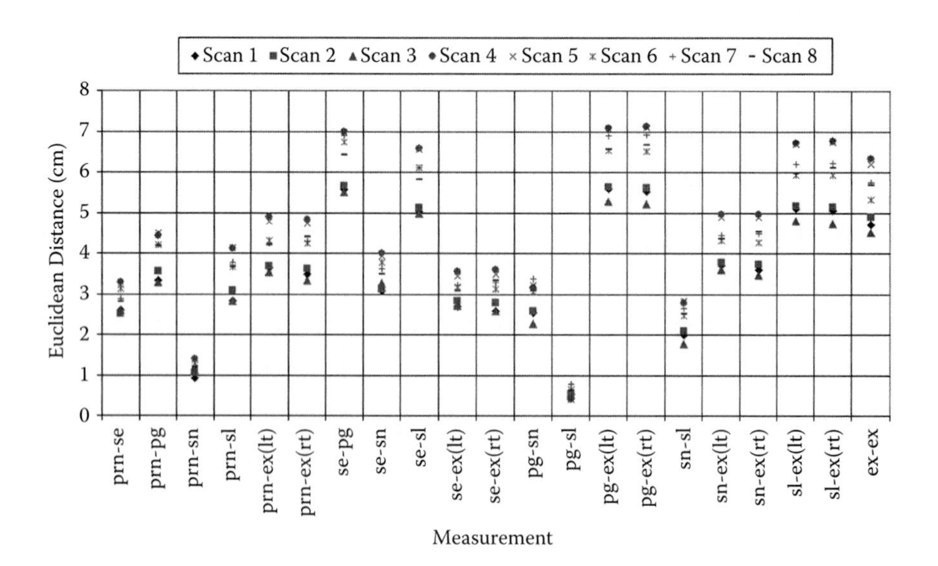

Figure 2.21 Chart showing minimum, maximum, and average pairwise distances between landmarks collected from scanned images at the 60,000 polygon resolution setting.

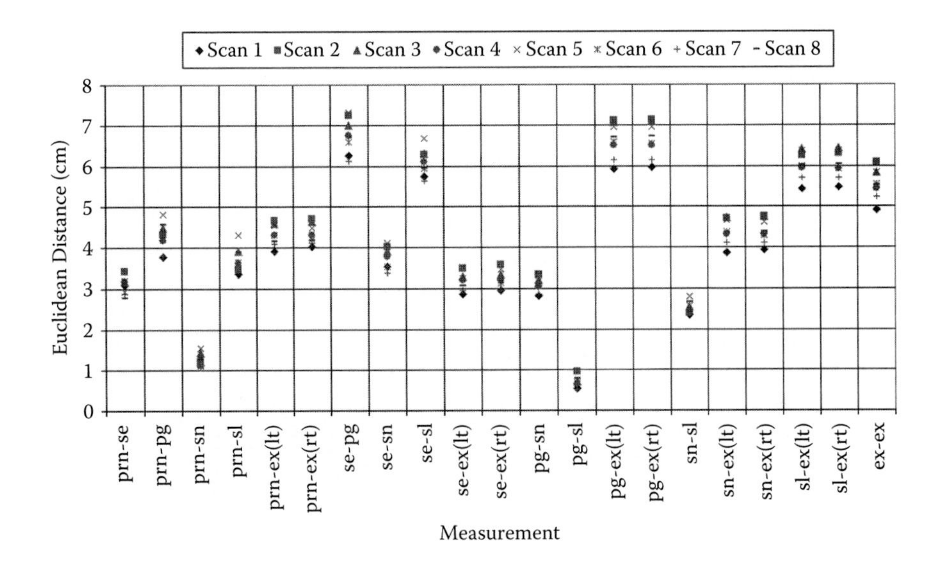

Figure 2.22 Chart showing minimum, maximum, and average pairwise distances between landmarks collected from scanned images at the 80,000 polygon resolution setting.

Maximum variation appears to differ between scanning resolutions. At 20,000 polygons (Figure 2.19), maximum variation is 13 mm (sl-ex r). At 40,000 polygons, an anomaly in the third scan has resulted in outliers in the chart (Figure 2.20). Maximum variation is about 14 mm (sl-ex r). At 60,000 polygon resolution (Figure 2.21), maximum variation is about 20 mm (sl-ex r), and many exceed 15 mm. Maximum variation in the 80,000 polygon resolution scans (Figure 2.22) is about 12 mm (se-pg and pg-ex l).

2.3.4 Discussion of Geometrix Scanner Performance

Differences in pairwise distance variation found at different polygon resolutions and in different scans are attributed to errors in scaling of the 3D surface and to anomalies in surfaces described in the assessment of image quality. Figure 2.23 shows an example of scaling difference in two images both captured at the 60,000 polygon resolution. The observed number of polygons recorded in each scan does not agree precisely with the resolution setting.

A number of anomalies were identified that have the potential to introduce random and systematic error in anthropometric measurements when they occur at landmark sites. These include scaling errors, localized anomalies, and texture map misalignment. Movement of the subject is again revealed as a source of anomalies in the 3D image.

Figure 2.23 Two images of different scales, both captured at 60,000 polygon resolution.

2.4 3dMDface System and Comparison between Scanners

The 3dMDface System scanner is shown in Figure 2.24.

The 3dMDface System scanner also utilizes stereophotography to capture 3D surface geometry and a texture map.

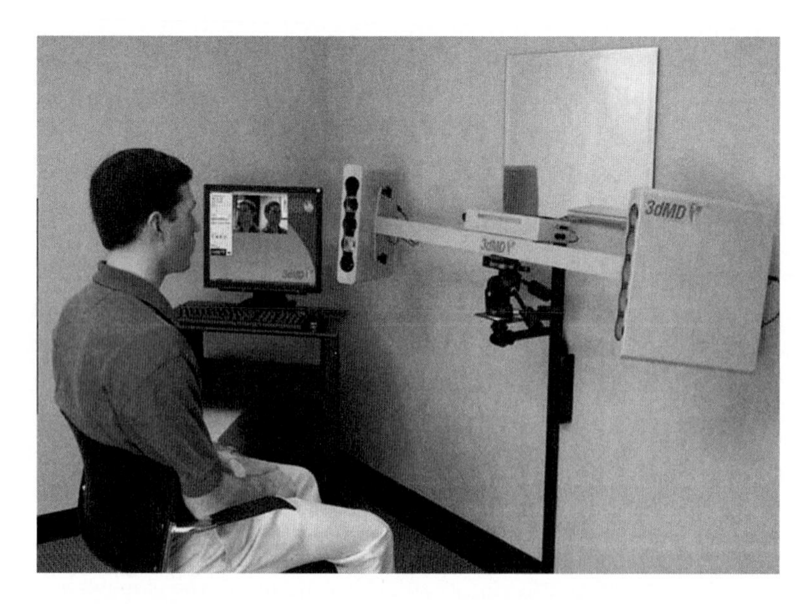

Figure 2.24 3dMDface System scanner. (From www.3dmd.com.)

2.4.1 Method of Assessment

Performance of the 3dMDface System scanner was evaluated in two ways: first, via a subjective assessment of 3D image quality and, second, via an empirical comparison of measurements collected from 3D scanned images with those collected directly from live subjects.

Single 3D image files of three subjects were captured with the 3dMDface System scanner in 3ds format following the manufacturer's instructions. Each 3ds file was then imported into 3ds Max and seven anthropometric landmarks were placed on each wireframe model in 3ds Max, namely the pronasale (prn), sellion (se), pogonion (pg), subnasale (sn), sublabiale (sl), and exocanthion left (ex l) and right (ex r) landmarks (see Figure 2.2).

The Euclidean distances for every combination of two landmarks (21 pairwise distances in total) were measured for each of the three files representing the three live subjects.

In order to assess accuracy of 3D geometry, pairwise distance measurements collected from the scanned files were compared with measurements collected with calipers.

The three subjects scanned using the 3dMDface System scanner had also been scanned (at different data collection locations) with the Cyberware 3030PS Head and Neck Scanner and the Geometrix FaceVision FV802 Series Biometric Camera, and had also been measured directly with digital calipers. These additional datasets permitted a limited direct comparison to be undertaken between all three scanners and the live subjects.

2.4.2 Assessment of Image Quality

Texture map misalignment with the 3D wireframe appears to be a minor or nonexistent anomaly with the 3dMDface System scanner (Figure 2.25), and random localized anomalies were not observed in this small sample.

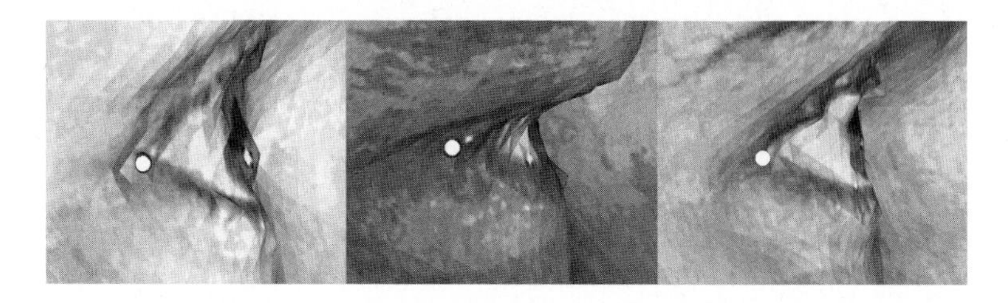

Figure 2.25 Position of the exocanthion for the three subjects showing good texture map alignment.

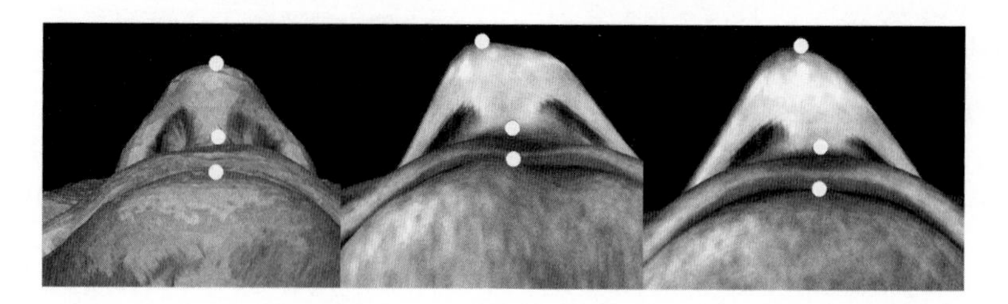

Figure 2.26 Position of the nasal tip for subject 2. Left: 3dMDface System; center and right: Geometrix FaceVision FV802 Series Biometric Camera.

Figures 2.26 and 2.27 show the geometry of the nose, comparing the 3dMDface System scanner and the Geometrix FaceVision FV802 Series Biometric Camera.

2.4.3 Comparison of Pairwise Measurements

Results of pairwise measurements collected from 3D images with each scanner and from live subjects with calipers are shown in Tables 2.1 to 2.3. Results are plotted in charts shown in Figures 2.28 to 2.30.

Tables 2.1 to 2.3 and Figures 2.28 to 2.30 consistently show that greatest variation in pairwise distance measurements arise from the Geometrix FaceVision FV802 Series Biometric Camera. The 3dMDface System scanner is consistently the most precise. Precision does not necessarily translate into accuracy in comparison with caliper measurements, however, where scanner performance appears to vary from subject to subject, and no clear leader emerges. This comparison is complicated by the error inherent in caliper measurement (see Discussion and Summary).

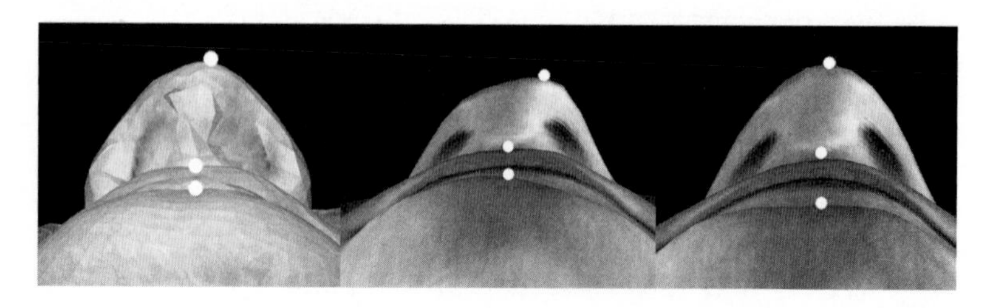

Figure 2.27 Position of the nasal tip for subject 3. Left: 3dMDface System scanner; center and right: Geometrix FaceVision FV802 Series Biometric Camera.

Table 2.1 Pairwise Distances between Landmarks in cm for Subject 1

Landmark Pair	Measurement by Scanner						Caliper Measurement
	3dMDface		Cyberware		Geometrix		
	Average	Range	Average	Range	Average	Range	
prn-se	5.169	0.084	5.112	0.122	5.165	0.446	5.111
prn-pg	6.664	0.040	6.480	0.125	6.749	0.185	6.454
prn-sn	1.996	0.051	1.993	0.091	2.115	0.282	2.075
prn-sl	5.849	0.047	5.649	0.091	6.022	0.262	5.648
prn-ex l	7.476	0.090	7.303	0.374	6.922	0.933	7.201
prn-ex r	7.334	0.132	7.271	0.183	7.048	0.581	7.119
se-pg	10.594	0.157	10.345	0.226	10.727	0.629	10.465
se-sn	5.893	0.116	5.799	0.152	6.131	0.713	5.687
se-sl	9.610	0.157	9.315	0.095	9.885	0.537	9.146
se-ex l	5.307	0.068	5.085	0.258	5.018	0.678	5.181
se-ex r	4.992	0.124	4.949	0.380	5.104	0.529	5.127
pg-sn	4.935	0.064	4.770	0.159	4.837	0.505	4.999
pg-sl	1.000	0.044	1.053	0.167	0.851	0.494	1.323
pg-ex l	10.850	0.093	10.566	0.300	10.497	0.189	10.792
pg-ex r	10.768	0.075	10.566	0.201	10.542	0.520	10.536
sn-sl	4.033	0.042	3.835	0.065	4.052	0.369	3.786
sn-ex l	7.425	0.078	7.218	0.331	7.030	0.458	7.095
sn-ex r	7.261	0.090	7.184	0.111	7.163	0.351	6.821
sl-ex l	9.962	0.107	9.626	0.238	9.738	0.339	9.477
sl-ex r	9.866	0.115	9.618	0.115	9.783	0.588	9.282
ex-ex	9.191	0.081	8.951	0.439	9.024	0.809	9.400

2.5 Discussion and Summary

Common sources of error in anthropometry—accruing from both manual and digital sources—are widely known (Farkas 1994, Bush and Antonyshyn 1996, Coward et al. 1997, El-Hussuna 2003). Key sources include improper identification of landmarks, improper positioning or measuring technique, and inadequate measuring equipment. As confirmed in this investigation, motion artifacts and head position or inclination may also influence surface geometry and, hence, landmark position. Comparison between measurements collected using calipers from living individuals and those collected from digital analogs encounter particular problems. Identifying landmarks on the face and on an image on the screen are different processes. Careful identification of certain landmarks may require palpation of the facial surface—a process clearly impossible on the screen. Although facial expression and other biological variation is

Table 2.2 Pairwise Distances between Landmarks in cm for Subject 2

	Measurement by Scanner						
Landmark Pair	3dMDface		Cyberware		Geometrix		Caliper Measurement
	Average	Range	Average	Range	Average	Range	
prn-se	4.762	0.038	4.663	0.102	4.167	0.360	4.671
prn-pg	7.056	0.004	6.931	0.177	7.102	0.286	7.260
prn-sn	1.872	0.008	1.956	0.048	2.040	0.343	2.089
prn-sl	5.797	0.004	5.730	0.182	5.870	0.452	5.740
prn-ex l	7.140	0.158	6.971	0.185	6.629	0.900	7.253
prn-ex r	7.203	0.206	7.131	0.344	6.624	0.384	7.202
se-pg	10.902	0.044	10.667	0.208	10.462	0.688	10.917
se-sn	5.449	0.047	5.361	0.092	5.226	0.299	5.424
se-sl	9.383	0.045	9.188	0.232	9.068	0.431	9.229
se-ex l	5.203	0.197	4.993	0.250	5.013	0.186	5.683
se-ex r	5.195	0.226	5.139	0.197	5.236	0.385	5.695
pg-sn	5.564	0.007	5.395	0.246	5.323	0.512	5.710
pg-sl	1.579	0.002	1.554	0.077	1.424	0.391	1.877
pg-ex l	11.177	0.035	10.891	0.275	10.572	0.526	11.025
pg-ex r	11.262	0.081	10.949	0.349	10.684	0.554	11.137
sn-sl	4.152	0.008	4.022	0.219	3.993	0.174	3.979
sn-ex l	7.051	0.095	6.829	0.200	6.618	0.487	7.101
sn-ex r	7.109	0.174	6.949	0.266	6.812	0.300	7.171
sl-ex l	9.764	0.036	9.500	0.224	9.300	0.370	9.402
sl-ex r	9.849	0.086	9.548	0.283	9.422	0.300	9.686
ex-ex	9.317	0.249	8.858	0.087	9.066	0.271	10.069

recognized as a potential source of error in anthropometry, this investigation shows that surface properties like luster are a further potential source. Caliper measurement may involve other particular sources of error. The potential for motion error may be exacerbated when a sharp metal instrument is placed near the eyes and other sensitive tissues, and operators may themselves contribute to motion error. Differences between live and digital measurements involving the left (ex l) and right (ex r) exocanthions (see Tables 2.1 to 2.3) are possibly attributable to this cause. There also appear to be some systematic errors—possibly observer error—involving different positioning of the sublabiale (sl) when collected digitally and from live subjects.

In summary, the following findings were encountered that affect image quality:

- Motion artifacts
- Spikes due to tissue surface luster, particularly affecting the Cyberware 3030PS Head and Neck Scanner

Table 2.3 Pairwise Distances between Landmarks in cm for Subject 3

Landmark Pair	Measurement by Scanner						
	3dMDface		Cyberware		Geometrix		Caliper Measurement
	Average	Range	Average	Range	Average	Range	
prn-se	5.504	0.035	5.646	0.052	4.683	0.897	4.742
prn-pg	5.761	0.055	5.712	0.071	5.754	0.474	5.967
prn-sn	1.798	0.086	1.638	0.089	1.618	0.110	1.801
prn-sl	4.767	0.032	4.833	0.091	4.916	0.194	4.897
prn-ex l	7.073	0.136	6.833	0.201	6.637	0.527	6.730
prn-ex r	7.343	0.039	7.257	0.095	7.109	0.268	6.734
se-pg	10.538	0.081	10.640	0.057	9.820	0.509	10.079
se-sn	6.076	0.096	6.120	0.131	5.503	0.752	5.244
se-sl	9.370	0.061	9.578	0.122	8.879	0.862	8.681
se-ex l	5.440	0.031	5.324	0.044	5.310	0.271	5.322
se-ex r	5.377	0.153	5.514	0.232	5.467	0.410	5.512
pg-sn	4.502	0.031	4.581	0.100	4.395	0.370	4.929
pg-sl	1.187	0.044	1.091	0.084	0.954	0.525	1.391
pg-ex l	10.299	0.052	9.901	0.098	9.858	0.435	9.962
pg-ex r	10.217	0.060	10.034	0.100	9.897	0.335	9.889
sn-sl	3.371	0.077	3.573	0.049	3.492	0.120	3.551
sn-ex l	6.970	0.038	6.590	0.128	6.790	0.187	6.574
sn-ex r	6.913	0.083	6.810	0.079	6.898	0.222	6.417
sl-ex l	9.263	0.088	8.936	0.165	9.029	0.143	8.683
sl-ex r	9.165	0.074	9.067	0.089	9.071	0.283	8.668
ex-ex	9.082	0.130	8.726	0.201	9.052	0.159	9.239

- Localized anomalies in tissue surface affecting the Geometrix FaceVision FV802 Series Biometric Camera
- Surface geometry and texture map misalignment affects the Geometrix FaceVision FV802 Series Biometric Camera, but is not evident in the other two
- Localized distortion, including possibly systematic distortion of the position of the nasal tip, affecting the Geometrix FaceVision FV802 Series Biometric Camera

Each of these anomalies is a potential source of error in anthropometric landmark placement. Ridges in images collected from mannequins were also evident, affecting the Geometrix FaceVision FV802 Series Biometric Camera, but do not appear to arise from scans of living individuals. This observation is somewhat academic in that the Geometrix FaceVision FV802 Series Biometric Camera was developed to work on live faces, not mannequins.

Repeatability and accuracy of measurements derived from 3D images captured using the Cyberware 3030PS Head and Neck and 3dMDface

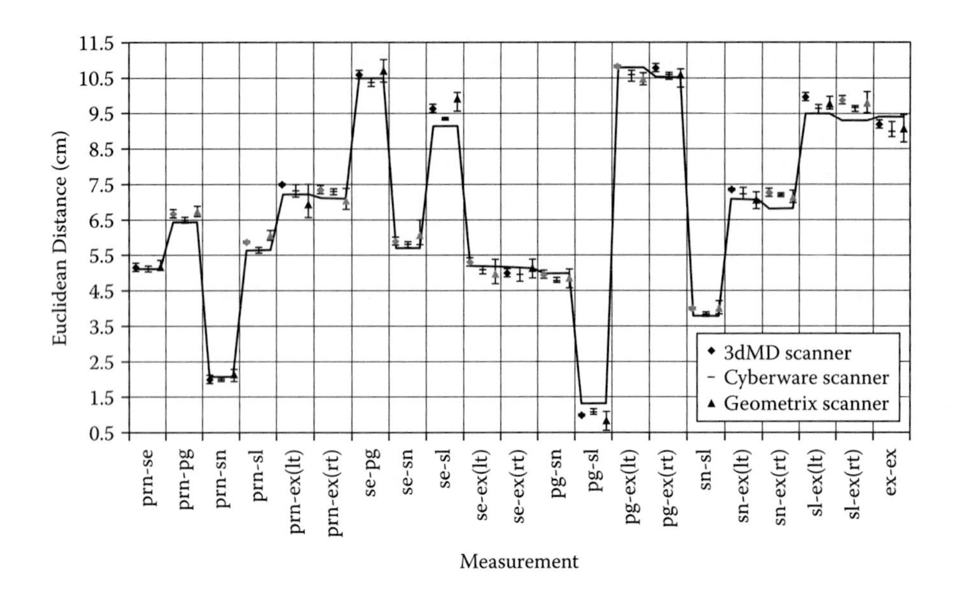

Figure 2.28 Chart showing minimum, maximum, and average pairwise distances between landmarks collected from three scanners (bars) and live measurements (line) for subject 1.

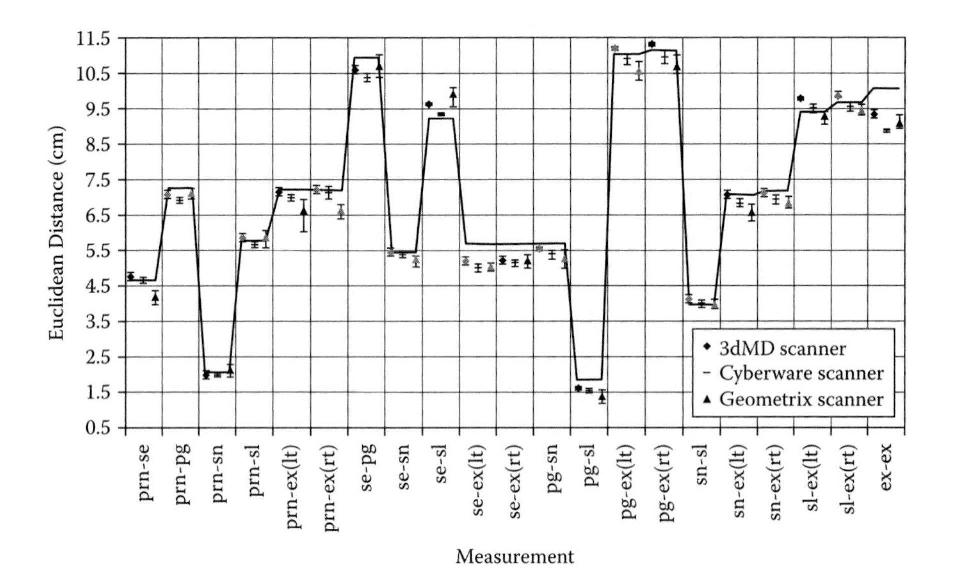

Figure 2.29 Chart showing minimum, maximum, and average pairwise distances between landmarks collected from three scanners (bars) and live measurements (line) for subject 2.

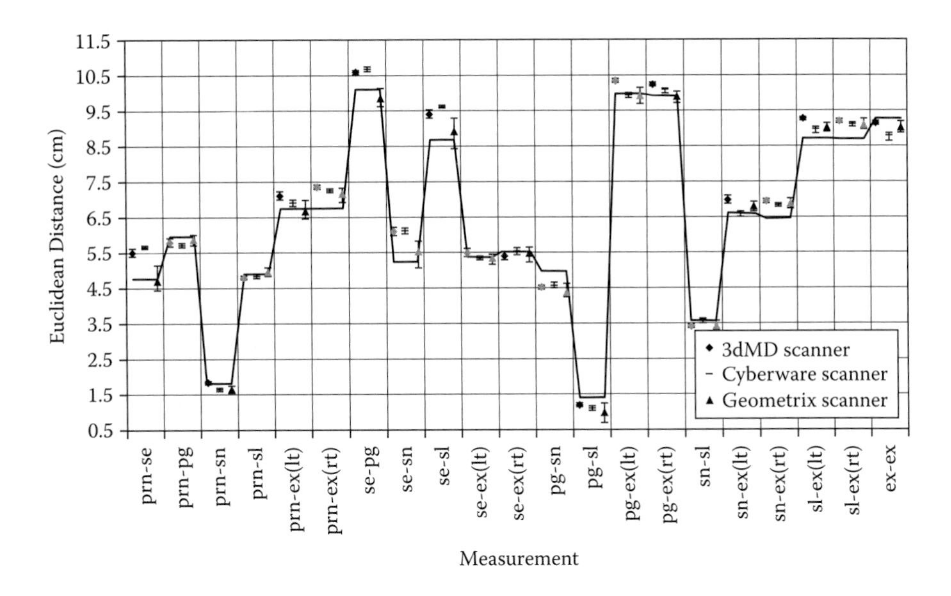

Figure 2.30 Chart showing minimum, maximum, and average pairwise distances between landmarks collected from three scanners (bars) and live measurements (line) for subject 3.

System scanners, assessed against caliper measurements collected from living individuals, were comparable to that reported in another study (Coward et al. 1997). The 3dMDface System scanner shows an impressive degree of precision in the small sample investigated, not necessarily manifested in greater accuracy than the other systems, however.

While the Geometrix FaceVision FV802 Series Biometric Camera appears to generate relatively poor quality 3D images of varying accuracy, it is important to note that the system of anthropometric measurement intended for use with this system is entirely separate. This system is considered in detail in the next chapter.

References

Bush, K., and O. M. Antonyshyn, 1996. Three-dimensional facial anthropometry using a laser surface scanner: Validation of the technique. *Plast. Reconstr. Surg.* 98(2): 226–35.

Coward, T. J., R. M. Watson, and B. J. Scott, 1997. Laser scanning for the identification of repeatable landmarks of the ears and face. *Br. J. Plast. Surg.* 50(5): 308–14.

El-Hussuna, A., 2003. Statistical variation of three dimensional face models. Masters diss., IT-University of Copenhagen. http://www.sumer.dk/Researches/thesis-02.pdf (accessed January 31, 2008).

Farkas, L. G., 1994. *Anthropometry of the head and face*, 2nd ed. New York: Raven Press.

Shape Variation in Anthropometric Landmarks in 3D

3

LUCY MORECROFT, NICK R.J. FIELLER,
IAN L. DRYDEN, AND MARTIN PAUL EVISON

Contents

3.1 Introduction

The Geometrix FaceVision® FV802 Series Biometric Camera offered the only commercially available portable 3D image scanner capable of capturing the face and ears, and offering a purpose-designed application tool—ForensicAnalyzer®—for precisely placing anthropometric landmarks in 3D. This system was, therefore, selected for a more extensive study of face shape variation in 3D, based on traditional anthropometric landmarks (Farkas 1994).

Provisional criteria were identified for a selection of landmarks suitable for use in research and practice in forensic facial comparison:

- Must be sufficiently visible in research subjects and forensic images
- Must be sufficiently variable to permit statistical estimation of a match or exclusion
- Must be amenable to collection of a large database sample
- Must be amenable to computerized statistical analysis
- Must be amenable to quantification of error from a variety of sources
- Variation due to error must not exceed that due to face shape variation

In this chapter, an investigation of anthropometric landmarks is presented in which their utility in distinguishing between individual subjects is assessed in comparison with their repeatability in placement in 3D. The result of the investigation is a final selection of 30 landmarks intended to optimally fulfill the requirements listed above, and to be used in the collection of a large database sample (as described in Chapter 4).

3.2 Facial Image Sample Used in the Investigation

In order to conduct a preliminary investigation of the utility of landmarks to distinguish between faces and the error associated with their placement on facial images in 3D, a pilot sample was prepared. The pilot sample consisted of 3D facial images of 35 different subjects captured using a Geometrix FaceVision FV802 Series Biometric Camera (ALIVE Tech, Cumming, GA). Of a possible 62 standard anthropometric landmarks (Figure 3.1 and Table 3.1), all were used except one. The trichion was abandoned at an early stage as it is frequently obscured by hair or rendered invisible by baldness. The remaining 61 landmarks were placed on each 3D image three times (repetitions 0, 1,

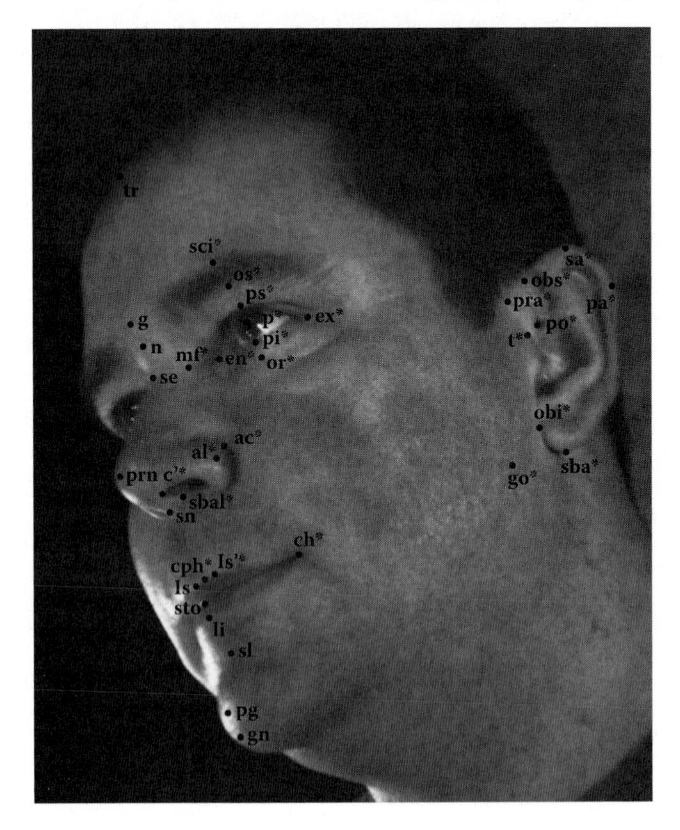

Figure 3.1 The 62 traditional anthropometric landmarks (after Farkas 1994). Note that the landmarks suffixed with an asterisk are the 25 bilateral landmarks, but only the left side of each is shown. For example the right exocanthion (ex r), is shown as ex*.

Table 3.1 Nomenclature for the 62 Traditional Anthropometric Landmarks Shown in Figure 3.1

Number	Name	Label
1	Glabella	g
2	Trichion	tr
3	Gonion left	go l
4	Gonion right	go r
5	Sublabiale	sl
6	Pogonion	pg
7	Gnathion	gn
8	Endocanthion left	en l
9	Endocanthion right	en r
10	Exocanthion left	ex l
11	Exocanthion right	ex r
12	Center point of pupil left	p l
13	Center point of pupil right	p r
14	Orbitale left	or l
15	Orbitale right	or r
16	Palpebrale superius left	ps l
17	Palpebrale superius right	ps r
18	Palpebrale inferius left	pi l
19	Palpebrale inferius light	pi r
20	Orbitale superius left	os l
21	Orbitale superius right	os r
22	Superciliare left	sci l
23	Superciliare right	sci r
24	Nasion	n
25	Subnasion	se
26	Maxillofrontale left	mf l
27	Maxillofrontale right	mf r
28	Alare left	al l
29	Alare right	al r
30	Pronasale	prn
31	Subnasale	sn
32	Subalare left	sbal l
33	Subalare right	sbal r
34	Alar crest left	ac l
35	Alar crest right	ac r
36	Highest point of columella prime left	c′ l
37	Highest point of columella prime right	c′ r
38	Crista philtri left	cph l
39	Crista philtri right	cph r

(continued)

Table 3.1 Nomenclature for the 62 Traditional Anthropometric Landmarks Shown in Figure 3.1 (Continued)

Number	Name	Label
40	Labiale superius	ls
41	Labiale superius prime left	ls′ l
42	Labiale superius prime right	ls′ r
43	Labiale inferius	li
44	Stomion	sto
45	Cheilion left	ch l
46	Cheilion right	ch r
47	Superaurale left	sa l
48	Superaurale right	sa r
49	Subaurale left	sba l
50	Subaurale right	sba r
51	Postaurale left	pa l
52	Postaurale right	pa r
53	Otobasion superius left	obs l
54	Otobasion superius right	obs r
55	Otobasion inferius left	obi l
56	Otobasion inferius right	obi r
57	Porion left	po l
58	Porion right	po r
59	Tragion left	t l
60	Tragion right	t r
61	Preaurale left	pra l
62	Preaurale right	pra r

and 2) by each of two operators (observers L and X) using ForensicAnalyzer (ALIVE Tech, Cumming, GA), generating 210 facial landmark datasets in total. Subjects were chosen, and their 3D images captured to ensure that all 61 landmarks were visible, as the subsequent statistical analysis would not have been possible if there are missing values in the data.

3.3 Analysis of Shape Variation and Observer Error

The methods of analysis were, generally, those of shape analysis (see Dryden and Mardia 1998), and were conducted by statistical programming in R (R 2008), utilizing a number of features of R, particularly generalized Procrustes analysis and principal components analysis (PCA). The sophisticated graphics features of R were valuable in visualizing the results of PCA, and Wilkes' lambda and Mahalanobis distance calculations.

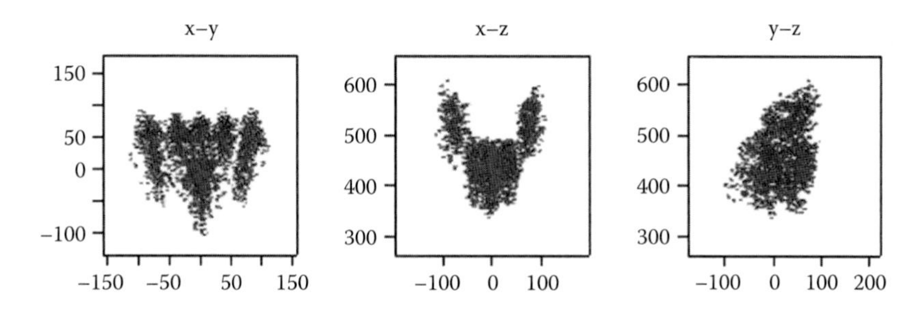

Figure 3.2 Orthographic plot of all 12,810 coordinates in the facial landmark sample.

Scatter plots were used to visualize the overall distribution of landmarks in the sample. Figure 3.2 shows a plot representing the entire sample in anterior, superior, and lateral views.

Procrustes registration was completed without scaling to enable direct comparison of the landmark datasets. This was undertaken via a principal components analysis (PCA) of the residual values calculated by subtracting the overall mean from the Procrustes-aligned data. Figure 3.3 shows a scatter plot of the overall distribution of landmarks in the sample after Procrustes registration.

The initial PCA indicated significant interobserver error. This is illustrated in Figure 3.4, where the distances between mean values for each landmark are depicted for the two observers. Candidate problem landmarks with large differences between the means of the two observers can be seen in the figure.

To assess the statistical significance of these differences and to establish a ranking of landmarks by observer error, Mahalanobis distances between observers were calculated for each landmark and subject in the nonregistered sample. This measure removes effects of correlation and scale. As there were only three observations of each landmark per observer, 1 was added to the variance in each case to make distances more stable. This has the effect that

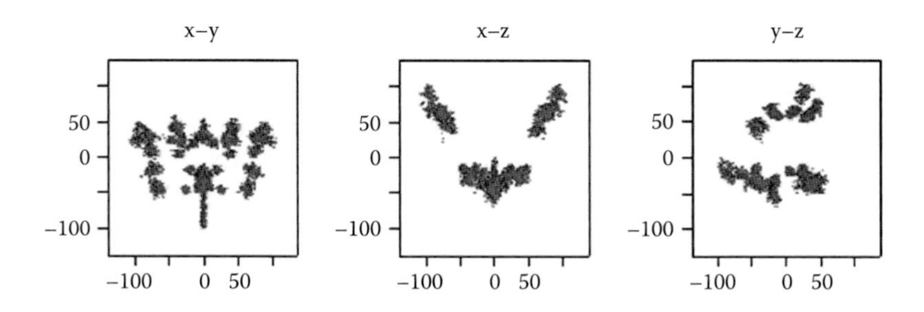

Figure 3.3 Orthographic plot of all 12,810 coordinates in the facial landmark sample after Procrustes registration without scaling.

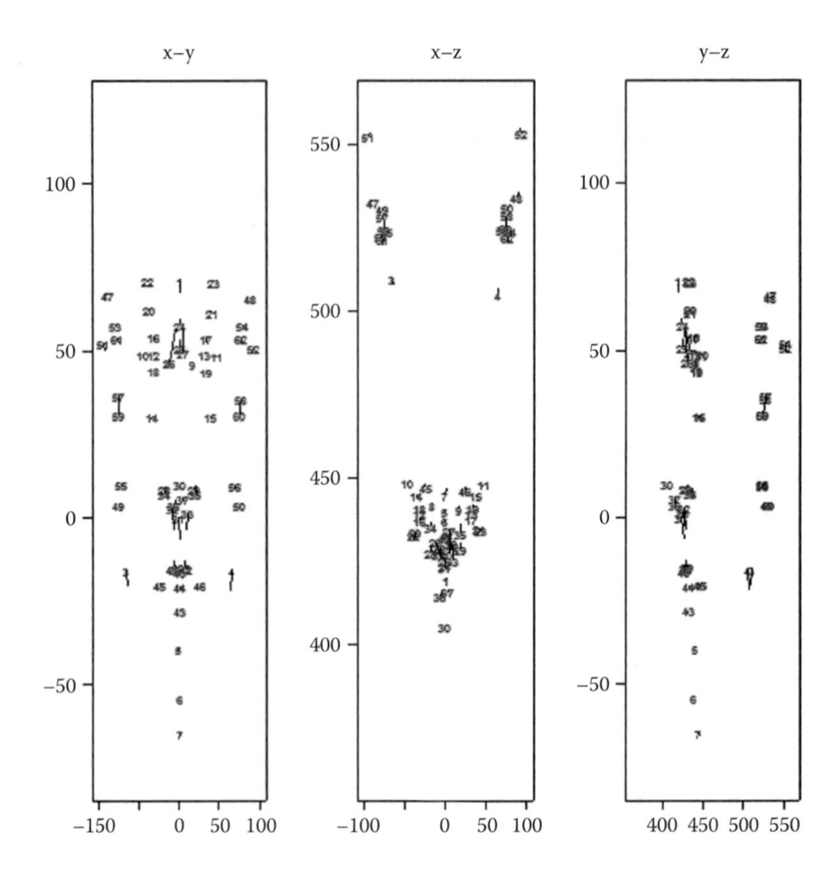

Figure 3.4 Orthographic plots, without Procrustes registration, of the mean landmark values for each of 61 landmarks placed by observer L with black vectors to the mean values for observer X.

only larger mean distances contribute to a large Mahalanobis distance. The results are shown in Figure 3.5.

Large distances are frequently evident at landmarks 26 and 27 (maxillofrontales), 31 (subnasale), 32 and 33 (subalares), 57 and 58 (porions), and 59 and 60 (tragions). Landmarks 3 and 4 (gonions) also exhibit slightly larger distances somewhat frequently. These findings appear to confirm the observations derived from comparing Euclidean distances between means (Figure 3.4).

Inconsistencies between observers in placing landmarks 26, 27, 31 to 33, and 57 to 60 continued despite quite specific efforts to remedy them by training and review. The maxillofrontales (26 and 27), in particular, appear to be potentially idiosyncratic to the observer and the others appear to present frequent ambiguities. A further PCA was undertaken with these landmarks excluded. The results obtained with the remaining 52 landmarks are shown in Figure 3.6, and no further significant distinction between observers was evident.

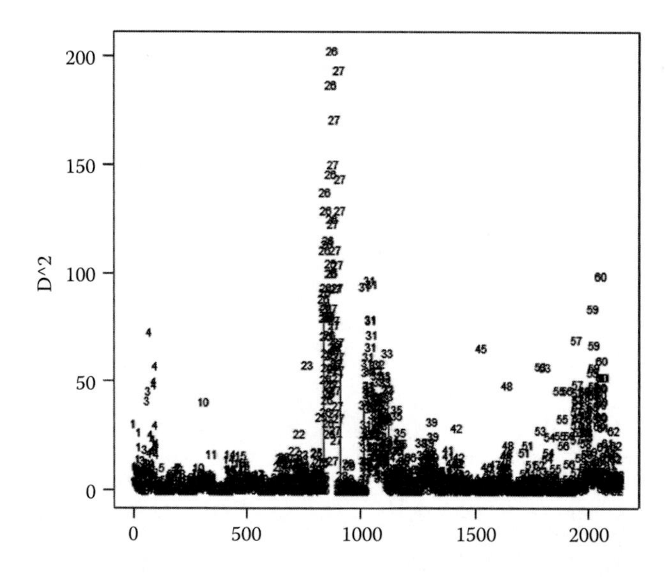

Figure 3.5 Mahalanobis distances between observers for each of 61 landmarks for each subject with the median distance for all subjects plotted as a gray line.

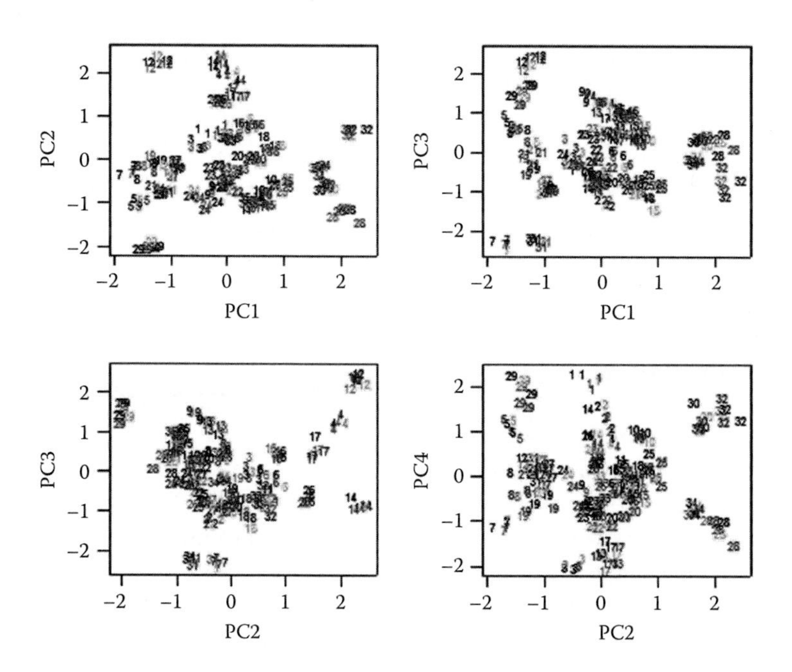

Figure 3.6 Results of a PCA analysis of the 3D facial image sample showing the first four PCs, permitting comparison of landmark datasets by subject (subjects 1-35), observer and repetition (represented by six gray scale values in the Figure). No significant influence of observer error is evident.

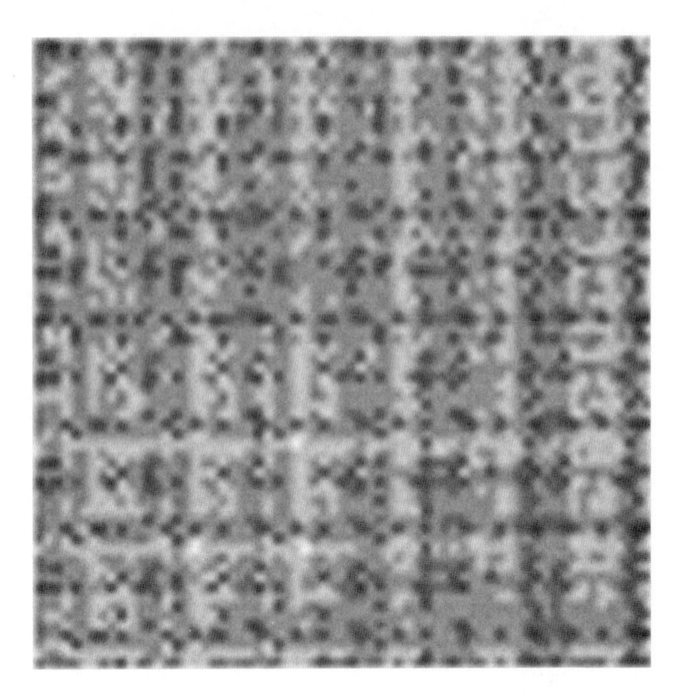

Figure 3.7 Log Euclidean distance matrix between PC scores for 52-landmark datasets in sets of six per subject. The matrix is constructed by face dataset in the sequence L0, L1, L2, X0, X1, X2 for each subject, where L and X are the observers and 0, 1 and 2 are the repetitions. Lighter shades indicate smaller distances.

In the absence of a significant influence of observer error, the power of landmarks to distinguish between faces was addressed. In order to provide a preliminary assessment, log Euclidean distances between PC scores for face datasets were calculated for the remaining 52 landmarks and are displayed as a matrix in Figure 3.7.

Distances are displayed in gray scale values where lighter shades illustrate smaller distances and darker shades indicate larger ones. The matrix is constructed by face dataset in the sequence L0, L1, L2, X0, X1, X2 for each subject, where L and X are the observers and 0, 1, and 2 are the repetitions.

The matrix can be used to test a simple hypothesis. If landmark-based shape variation between subjects' faces is significant and greater than that due to observer error, a clear grid pattern can be predicted in the matrix: since each set of six face dataset arises from the same subject, a light shade is expected for each set due to a small Euclidean distance between them, followed by a dark shade as the next dataset—from a different subject—is encountered. The figure supports this assumption.

A further assessment is offered by the results of a cluster analysis constructed using Ward's method (Figure 3.8) for the remaining 52 landmarks,

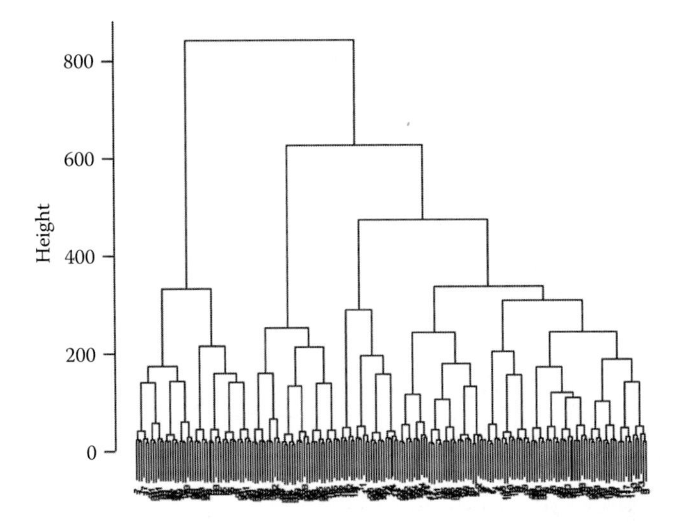

Figure 3.8 Dendrogram showing clustering of facial image landmark datasets. The datasets segregate uniformly into the anticipated six datasets per subject.

which shows uniform clustering of all 35 subjects' six landmark datasets, further confirming the potential of anthropometric landmarks to distinguish between faces in 3D.

Anthropometric landmark-based facial comparison must inevitably be based on a statistical study of shape variation utilizing a large study sample. An optimal subset of landmarks is, therefore, desirable for use in accumulating a large database of facial image landmark datasets.

In order to rank landmarks for their power to distinguish between landmark datasets, Wilks' lambda (Λ) was calculated for each landmark using Procrustes-registered data. Landmarks, which vary greatly between face datasets relative to between observers, have high values of $-\log \Lambda$. The values of $-\log \Lambda$ are shown in the three orthogonal views in Figure 3.9, plotted on

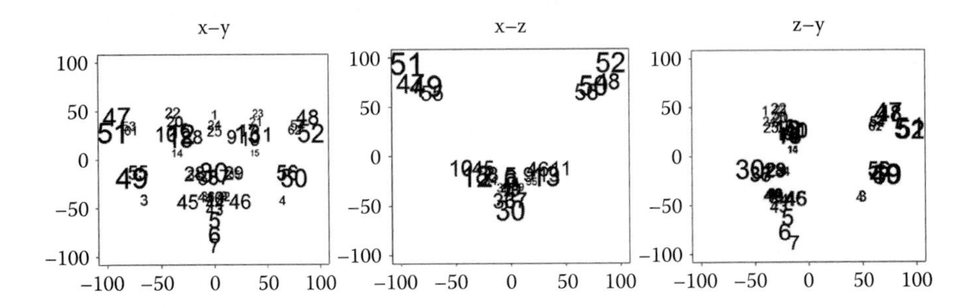

Figure 3.9 Plots of $-\log \Lambda$ for each of 52 landmarks represented as the three orthogonal views plotted on the overall mean face dataset, where the size of the landmark number is proportional to $-\log \Lambda$.

Table 3.2 Ranking of Landmarks by Power to Distinguish between Face Datasets (Read from Left to Right, Then Top to Bottom)

49	**51**	**47**	**30**	**50**	**12**	**52**	**6**	**13**	**5**
18	**48**	**56**	**10**	**46**	7	**37**	**45**	**55**	**11**
36	16	**8**	**19**	**9**	28	**40**	29	**44**	**43**
17	3	20	53	41	22	54	**25**	**1**	4
57	60	62	58	42	21	**38**	34	61	24
59	23	14	**39**	15	35	31	33	32	27
26									

Note: Retained landmarks are in bold text.

the overall mean face dataset values for the sample, where the size of the landmark number is proportional to $-\log \Lambda$.

Tables 3.2 and 3.3 provide rankings by $-\log \Lambda$ (see Figure 3.9) and Mahalanobis distances (Figure 3.5) for all 61 landmarks, indicating the power of landmarks to distinguish between face datasets and associated observer error, respectively.

Finally, a discriminant analysis was undertaken to assess the performance of landmarks in classification. Figure 3.10 shows the results of training on one observer's data and testing against the other.

In order to arrive at an optimal set of landmarks from the possible 62 given in Table 3.1, 30 landmarks were excluded. The excluded landmarks and the reason for their exclusion are given in Table 3.4.

The remaining landmarks appear in bold type in Tables 3.2 and 3.3. The primary reason for rejection of landmarks was that they were error prone or not powerful in distinguishing between face datasets. Of the landmarks that appear in the top half of the ranking based on power to distinguish between face datasets (Table 3.2), only landmarks 7, 16 and 17, and 28 and 29 were rejected. Landmark 7, the gnathion, was rejected because it is in a position under the chin that is difficult to locate (especially, it is anticipated,

Table 3.3 Ranking of Landmarks by Precision in Placement (Read from Left to Right, Then Top to Bottom)

30	**8**	**18**	**37**	28	**6**	**12**	**55**	**50**	**13**
49	**45**	**43**	**56**	**5**	**9**	**46**	**40**	29	**36**
19	**44**	7	54	53	23	24	**47**	41	57
21	**48**	16	**51**	22	17	20	34	14	**52**
11	15	**10**	42	61	62	**25**	58	3	**1**
38	35	59	**39**	60	4	32	33	31	27
26									

Note: Retained landmarks are in bold text.

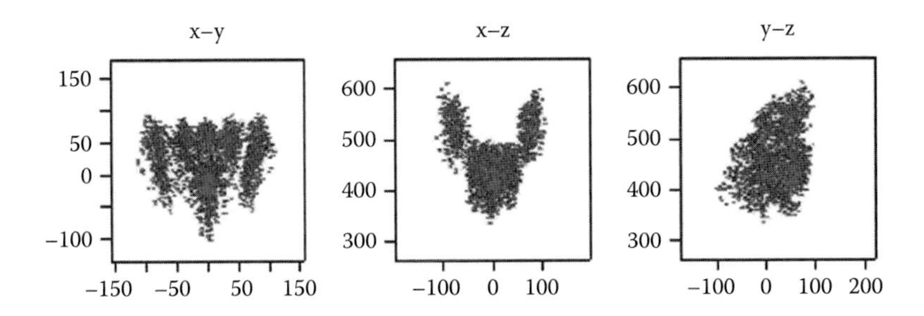

Figure 3.10 Orthographic plot of all 6,720 coordinates in the 32 landmark sample.

in security camera images) and because a close and better alternative is available in landmark 6, the pogonion. Landmarks 16 and 17 (palpebrale superius, left and right) were rejected because they are dependent on lack of closure of the eye, and are therefore unreliable and of uncertain value in comparison. Although apparently valuable, landmarks 28 and 29 (alare, left and right) were rejected as they were regarded by the operators as too time consuming to place accurately in a large database. As an alternative, adjacent landmarks 38 and 39 (crista philtri, left and right), were retained initially.

Table 3.4 List of Excluded Landmarks

Landmark	Landmark Name	Reason for Exclusion
2	Trichion	Frequently not visible
26, 27	Maxillofrontale (left and right)	Observer error (see Figure 3.5)
31	Subnasale	Observer error (see Figure 3.5)
32, 33	Subalare (left and right)	Observer error (see Figure 3.5)
57, 58	Porion (left and right)	Observer error (see Figure 3.5)
59, 60	Tragion (left and right)	Observer error (see Figure 3.5)
3, 4	Gonion (left and right)	Poorly ranked for power and consistency
7	Gnathion	Poorly ranked for power and consistency
14, 15	Orbitale (left and right)	Poorly ranked for power and consistency
16, 17	Palpebrale superius (left and right)	Poorly ranked for power and consistency
20, 21	Orbitale superius (left and right)	Poorly ranked for power and consistency
22, 23	Superciliare (left and right)	Poorly ranked for power and consistency
24	Nasion	Poorly ranked for power and consistency
41, 42	Labiale superius prime (left and right)	Poorly ranked for power and consistency
53, 54	Otobasion superius (left and right)	Poorly ranked for power and consistency
61, 62	Preaurale (left and right)	Poorly ranked for power and consistency
28, 29	Alare (left and right)	Too time consuming to place*

* Retained for pragmatic reasons in place of the crista philtri; see text.

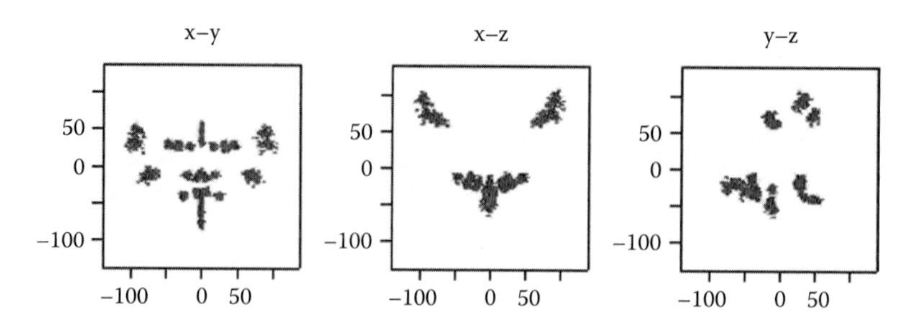

Figure 3.11 Orthographic plot of all 6,720 coordinates in the 32 landmark sample after Procrustes registration without scaling.

Landmarks 1, the glabella, and 25, the subnasion, were retained despite their middling to poor power to distinguish between face datasets, because they are readily visible on facial images and hence are of practical utility. Landmarks need to be visible to be useful in comparison.

Landmark 2, the trichion, was abandoned at the commencement of the study for the reasons given above.

Some of the rejected landmarks were middling in consistency of placement, but poor in discriminatory power and, hence, could be safely dispensed with.

In order to assess the effect of excluding these landmarks on relative variability, power to distinguish between face datasets and observer error, scatter plot, PC, matrix and cluster analyses were conducted on the remaining 32 landmarks in the sample, and $-\log \Lambda$, and Mahalanobis distances were calculated.

Unregistered and Procrustes-registered scatter plots are shown in Figures 3.10 and 3.11.

There is no evidence of observer error in the PCA (see Figure 3.12).

The Euclidean distance matrix analysis (Figure 3.13) shows more extensive segregation due to variation between face datasets relative to variation due to observer error.

Cluster analysis using Ward's method (Figure 3.14) again shows uniform clustering by subject.

Wilks' lambda (Figure 3.15) and Mahalanobis distances (Figure 3.16) provide an indication of the relative power of each landmark to distinguish between face datasets and associated observer error, respectively. Associated numerical values are given in Table 3.5.

Landmarks will inevitably vary in power to distinguish between faces and in susceptibility to observer error, and as poorly performing landmarks are excluded others will rise in the relative rankings. Landmark pair 34 and 35 (alar crests, left and right), performed poorly, however, as they had done relative to pair 28 and 29 (alares, left and right) in the study of 61 landmarks. Although the alares may take a little longer for observers to mark accurately, it was observed that they performed reasonably well on discrimination power

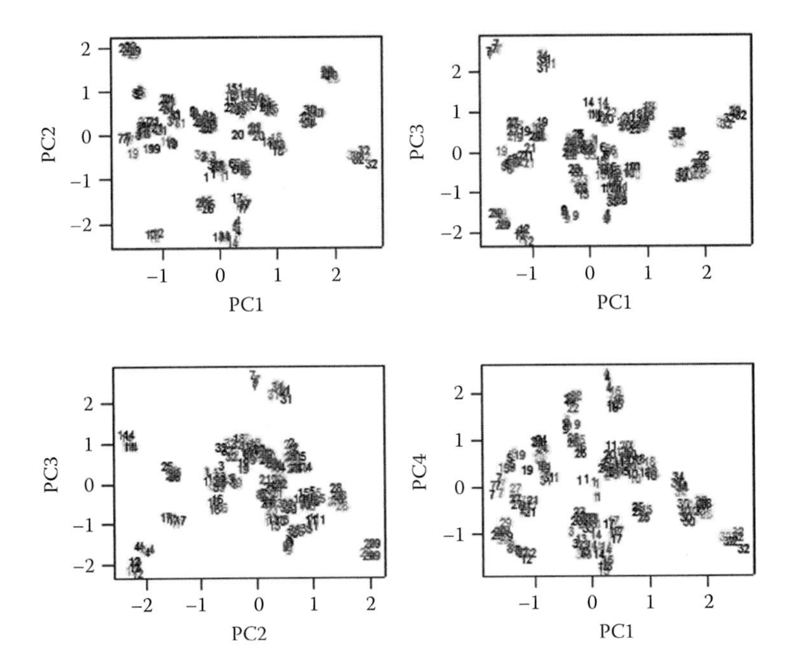

Figure 3.12 Results of a PCA analysis of the 3D facial image sample showing the first four PCs, permitting comparison of 32-landmark datasets by subject (subjects 1-35), observer and repetition (represented by six gray scale values in the Figure).

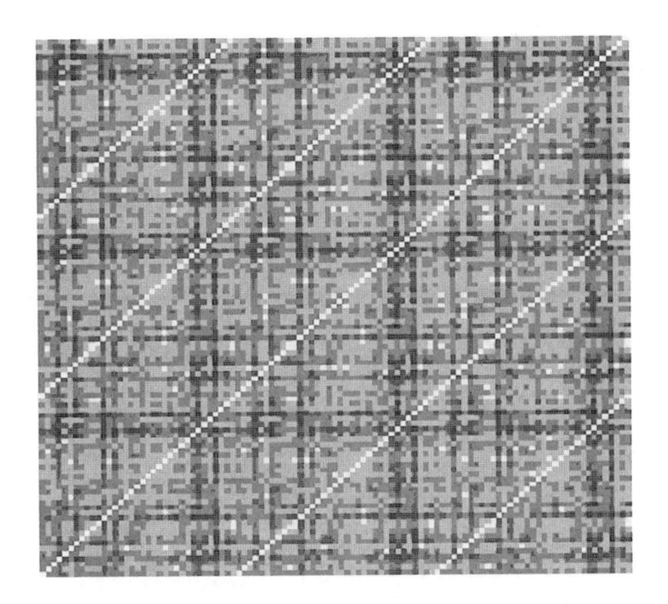

Figure 3.13 Log Euclidean distance matrix between PC scores for 32-landmark datasets in sets of six per subject. The matrix is constructed by face dataset in the sequence L0, L1, L2, X0, X1, X2 for each subject, where L and X are the observers and 0, 1 and 2 are the repetitions. Lighter shades indicate smaller distances.

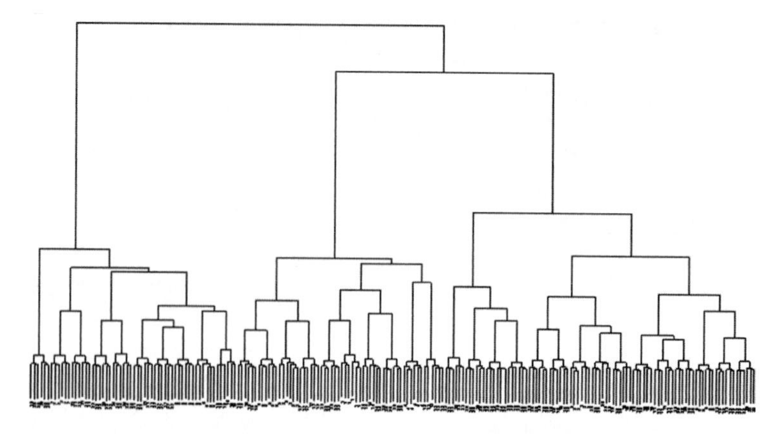

Figure 3.14 Dendrogram showing clustering of facial image landmark datasets. The datasets segregate uniformly into the anticipated six datasets per subject.

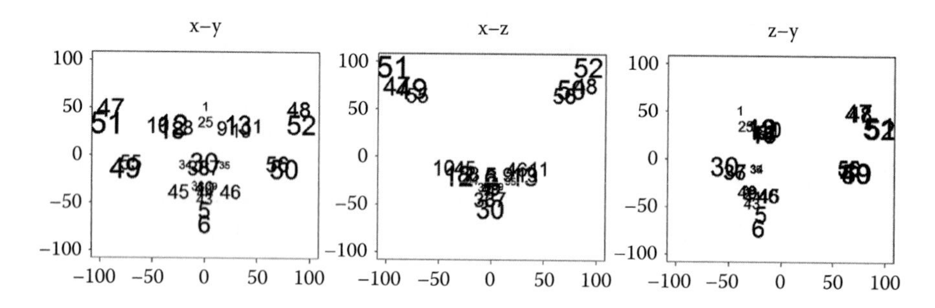

Figure 3.15 Plots of –log Λ for each of 32 landmarks represented as the three orthogonal views plotted on the overall mean face dataset, where the size of the landmark number is proportional to –log Λ.

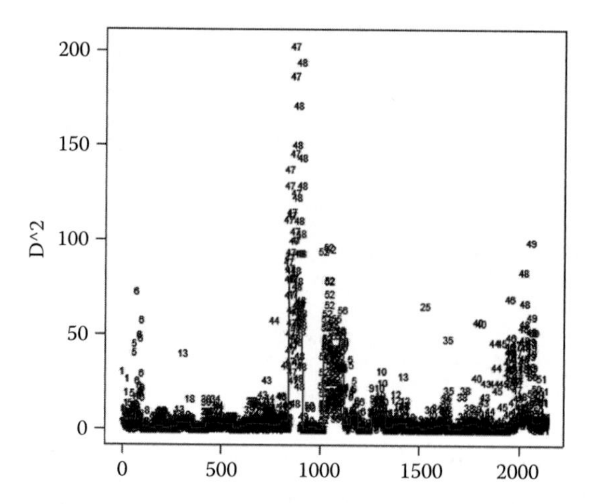

Figure 3.16 Mahalanobis distances between observers for each of 32 landmarks for each subject with the median distance for all subjects plotted as a gray line.

Table 3.5 Ranking of 32 Landmarks by $-\log \Lambda$ and Mahalanobis Distance

Ranking of Landmark by $-\log \Lambda$		Ranking of Landmark by Median Mahalanobis Distance	
Landmark Number	$-\log \Lambda$	Landmark Number	Median Mahalanobis Distance
51	11.083597	51	0.3273709
49	10.417496	11	0.5829767
12	10.401172	37	0.6907613
52	9.838848	49	0.7057878
13	9.568364	9	0.7170104
50	9.548721	19	0.7330481
30	9.500766	25	0.7772385
47	9.25138	8	1.0433323
6	8.647328	12	1.1271665
18	8.644748	50	1.461273
5	8.389833	38	1.6484862
10	7.717405	10	1.9891589
56	7.698128	44	2.2603939
48	7.673181	45	2.380513
36	7.371864	40	2.8510717
37	7.192176	35	2.9407126
55	7.137979	43	3.1160612
46	7.097231	36	3.1338087
45	7.054884	39	3.1707671
19	6.992865	30	3.2024407
8	6.985455	18	3.2842937
9	6.905168	34	3.3225045
11	6.798201	13	3.3623007
40	6.3902	46	4.8339315
44	5.425425	5	5.3425402
43	5.348737	1	5.4629451
25	4.881967	6	9.5283784
1	4.85232	55	28.4030799
34	4.378792	56	29.960089
38	4.281113	52	39.5901859
39	3.603852	48	64.4698875
35	3.534516	47	80.7393796

and accuracy in the pilot study, especially relative to the alar crests to which they are adjacent. Although using the alares would increase the landmarking time involved in collection of a large face database, the time factor would be negligible in a case-by-case comparison scenario. Since the alares are positioned on the sides of the nose and might be expected to feature in most facial images, it was decided to retain them at the expense of the alar crests, which were abandoned.

Conversely, landmarks 38 and 39 (crista philtri, left and right), which the operators recommended be retained because they are easily located, turned out to perform poorly in discriminating power and accuracy of placement in both the 61 landmark and 32 landmark studies, and where hence also dispensed with. The error rate in millimeters of the final 30 landmarks is given in Table 3.6.

Table 3.6 Error Rate in Millimeters of the Final 30 Landmarks

Landmark Number	Landmark	Label	Error Rate (mm)
27	Postaurale left	pa l	8.74
30	Otobasion inferius right	obi r	9.62
28	Postaurale right	pa r	9.63
4	Endocanthion left	en l	10.89
8	Center point of pupil left	p l	11.27
13	Alare left	al l	11.29
5	Endocanthion right	en r	11.45
17	Highest point of columella prime right	c′ r	11.63
21	Cheilion left	ch l	11.67
22	Cheilion right	ch r	11.75
9	Center point of pupil right	p r	11.76
19	Labiale inferius	li	11.80
14	Pronasale	prn	12.01
29	Otobasion inferius left	obi l	12.30
16	Highest point of columella prime left	c′ l	12.38
2	Sublabiale	sl	12.44
11	Palpebrale inferius right	pi r	12.92
6	Exocanthion left	ex l	12.99
25	Subaurale left	sba l	13.08
15	Alare right	al r	13.41
10	Palpebrale inferius left	pi l	13.46
3	Pogonion	pg	13.78
26	Subaurale right	sba r	13.80
18	Labiale superius	ls	13.97
1	Glabella	g	14.18
7	Exocanthion right	ex r	14.23
12	Subnasion	se	14.74
23	Superaurale left	sa l	15.87
24	Superaurale right	sa r	16.13
20	Stomion	sto	19.35

Rankings shown in Figures 3.14 and 3.15, and Tables 3.5 and 3.6, offer prioritization of landmarks in statistical software, equipment, and application tool design.

3.4 Conclusion

Figure 3.17 shows a comparison between the location of the original 62 and final 30 landmarks.

The remaining 30 landmarks continue to cover every region of the face and the face overall. By implication, facial comparison from any 2D pose should not be fundamentally compromised by the absence of landmarks because of the exclusions. The final 30 landmarks are proposed due to:

- Their visibility in research subjects and forensic images
- Their variability between individual subjects
- Their ease of placement in collection of a large database sample
- Their potential value in computerized statistical analysis
- A provisional associated quantification of error from a variety of sources
- The variation in repeatability shown not exceeding that due to face shape variation
- Optimal for collection in a large database, and for statistical and application tool development

Figure 3.17 Illustrative comparison of the original 62 and final 30 anthropometric landmarks (see also Figure 3.1).

References

Dryden, I. L., and K. V. Mardia, 1998. *Statistical shape analysis.* London: Wiley.
Farkas, L. G., 1994. *Anthropometry of the head and face*, 2nd ed. New York: Raven Press.
R., 2008. *The R project for statistical computing.* http://www.r-project.org (accessed June 30, 2009).

A Large Database Sample of 3D Facial Images and Measurements

4

MARTIN PAUL EVISON, LUCY MORECROFT, NICK R.J. FIELLER, AND IAN L. DRYDEN

Contents

4.1 Introduction

Tools for forensic facial comparison will inevitably rely on estimations of face shape frequency in the general population. In order to provide sample from which estimates of shape frequency can be derived, a database of facial images and measurements is essential. A database additionally allows questions of representativeness—such as by age, sex, and ancestry—to be addressed and, finally, it permits patterns of variation within the database itself to be understood with empirical confidence.

This chapter describes the collection of a large database of facial images and anthropometric landmark datasets in 3D. The investigation described in Chapter 3 served effectively as a pilot study in which an optimal selection of anthropometric landmarks was identified that could be used to collect measurements from a large quantity of facial images. The pilot investigation also allowed methods of training and quality assurance for operators to be developed.

4.2 Data Collection

The facial images—and measurements derived from them—were collected from public volunteers following ethical review (see Chapter 1). The primary site of data collection was the Magna Science Adventure Centre near Rotherham, South Yorkshire, England. Data was also collected at several other locations:

- Magna Science Adventure Centre, Rotherham, U.K.
- Biometrics 2004 Conference, London, U.K.
- Forensic Science Society Annual Meeting 2004, St. Neots, U.K.
- Sheffield Medico-Legal Centre, Sheffield, U.K.
- National Fingerprint Conference, Leicester, U.K.

The manufacturer's instructions were followed in the use of the Geometrix FaceVision® FV802 Series Biometric Camera. A standard operating procedure—in the form of the landmarking reference manual (see Appendices)—was produced for instruction and reference for landmarking operators. Two teams of graduate students were recruited, the first to collect the facial images and the second to place the anthropometric landmarks. Both teams were trained to use the equipment and follow the manufacturer's instructions and landmarking reference manual. The statistical analyses described in Chapter 3 were used as a means of quality assurance; namely, to ensure that the landmarking operator was not a source of significant interobserver or intraobserver error.

Data was collected from healthy volunteers of over 14 years of age. Members of the public who decided to volunteer were asked to complete a biographic information form, and complete and sign a consent form (see Appendix A). For volunteers aged 14 to 16 years, a parent or guardian was asked to sign the consent form. A Geometrix FaceVision FV802 Series Biometric Camera (ALIVE Tech, Cumming, GA) was used to collect facial images in 3D. A Cyberware® 3030PS Head and Neck Scanner (Cyberware, Monterey, CA) was also used to collect a subsidiary comparative database of 3D images. The biographic information and consent form information was keyed into two Microsoft® Office 2003 Excel spreadsheets: one containing biographic records, and the other containing consent records. The information was input twice in order to provide comparative verification of correct data entry. Both were password protected and kept in locked locations. On completion of data collection, the consent spreadsheet was secured to CD and kept in a separate locked location, with the hard drive copy deleted.

The consent spreadsheet was kept in a separate location to reduce the potential for a breach of confidentiality. It was necessary to retain the consent information for cross reference purposes, should a volunteer wish to withdraw from the project—and have their data deleted—or in case the volunteer wished to be contacted to be kept informed about the research.

Table 4.1 List of Fields in Each Biographic Record

Field Name	Data Type	Number of Characters	Restrictions	Occurrences in Record
Age	Number	2		1
Sex	Character	1	"M" or "F"	1
Ancestry code	Number	2	"01"–"15" (see Table 4.2)	1
Ancestry description	Character	Variable (14)		1
Relative	Block	Variable (6)	See Table 4.3	Variable
Scanner	Number	1	"1"–"3"[a]	1
Location	Number	1	"1"–"5"[b]	1
Date	Number	6	Format: ddmmyy	1
Key	Number	5	Unique	1
Scanner operator	Number	2	"01"–"10"	1

Note: The field's scanner, location, date, key, and operator are noted on the consent form; the concatenation of date and key provides a single unique key for each record.

[a] "1": Geometric; "2": Cyberware; "3": Both Geometrix and Cyberware.

[b] See list of locations in text above.

The fields recorded in the biographic and consent spreadsheets from the biographic information and consent forms are given in Tables 4.1 to 4.4, respectively.

The volunteer's image was captured by the operator, who provided them with their 3D image on diskette in VRML format as a souvenir (see Figure 4.1).

Table 4.2 Key to Ancestry Code in Biographic Record (see Figure 4.1)

Ancestry Code	Description	Description Option
01	White British	
02	Any other White background	Yes
03	White and Black Caribbean	
04	White and Black African	
05	White and Asian	
06	Any other Mixed background	Yes
07	Indian	
08	Pakistani	
09	Bangladeshi	
10	Any other Asian background	Yes
11	Caribbean	
12	African	
13	Any other Black background	Yes
14	Chinese	
15	Any other	Yes

Table 4.3 List of Fields in Each Relative Block of Biographic Record (see Figure 4.1)

Field Name	Data Type	Number of Characters	Restrictions	Occurrences in Block
Relative name	Character	Variable (17)		1
Relative date of birth	Number	Variable (6)	Format: ddmmyy	1
Relative relationship	Character	Variable (8)		1

Table 4.4 Descriptions in Each Record from the Consent form

Field Name	Data Type	Number of Characters	Restrictions	Occurrences in Record
Family name	Character	Variable (27)		1
First name	Character	Variable (27)		1
Date of birth	Number	Variable (6)	Format: ddmmyy	1
Email address	Character	Variable (28)	Format: email address	1

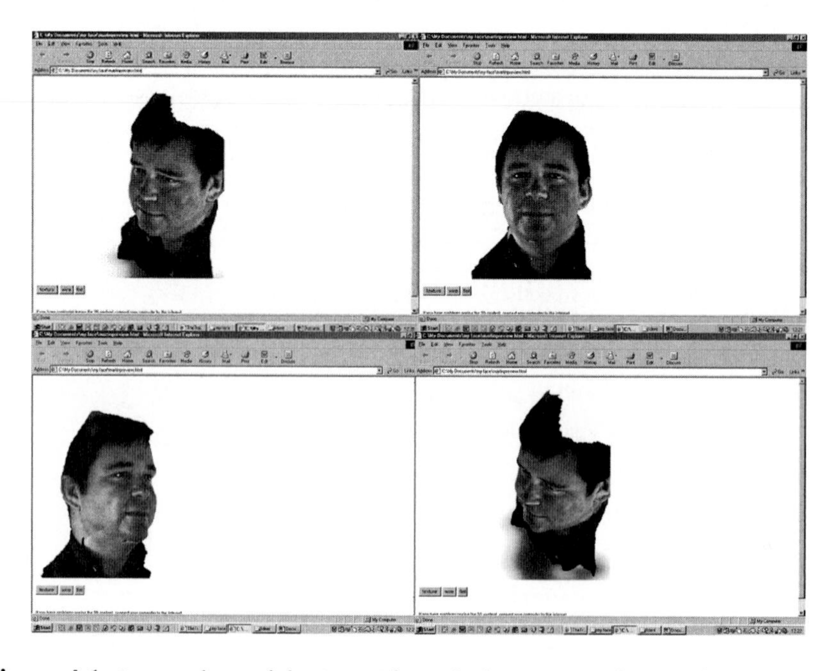

Figure 4.1 Screen shots of the 3D surface of a face captured using the Geometrix FaceVision FV802 Biometric Camera and viewed in Internet Explorer® with the Viewpoint Media Player.

Table 4.5 Description Fields in Each Record of the Landmark Spreadsheet

Field Name	Data Type	Number of Characters	Restrictions	Occurrences in Record
Date	Number	Variable (6)	Format: ddmmyy	1
Key	Number	Variable (5)	Unique	1
Landmarking	Number	Variable (1)	"1" or "2"	1
Landmarking operator	Number	Variable (1)	"1"–"6"	1
Coordinate (mm)[a]	Block	Variable		

Note: The concatenation of date, key, and landmarking provides a single unique key for each record, which represents one landmarking of each subject's face.

[a] The coordinate block consists of 30 × 3 cells of each spreadsheet record, comprising the x, y, and z coordinates of each landmark repeated for each of the 30 landmarks.

Image data was password protected and daily securities were taken to external hard disk drives, which were stored at separate, locked locations.

Landmarking of the Geometrix images was undertaken using Geometrix ForensicAnalyzer® (ALIVE Tech, Cumming, GA) to place craniofacial anthropometric landmarks. The anthropometric landmarks used are those 30 landmarks identified as optimal for the purpose in Chapter 2. Not all landmarks can be placed, however, as some are obscured—by head or facial hair, for example. Each 3D image was landmarked independently by two separate operators. The FaceVision software was also used to generate 3D surface images of each face in 3ds (Autodesk 3ds MAX®, San Rafael, CA), DXF (Autodesk AutoCAD®, San Rafael, CA) and VRML (Hartman and Wernecke 1996) formats.

Raw landmark data was transferred from the XML format files generated by ForensicAnalyzer to Microsoft Office Excel 2003 spreadsheets using a simple software application. The description of the fields in each record of the landmark spreadsheet is given in Table 4.5.

4.3 Description of the Database

Data collection and landmarking was completed in a 16-month period between December 2003 and May 2005. Upon completion, the Geometrix database included images of 3115 volunteers. The distribution according to age and ancestry is given in Table 4.6 and according to age and sex is given in Table 4.7.

The Geometrix database comprises 3115 folders, each containing eight JPEG images corresponding to the eight cameras used by the Geometrix

Table 4.6 Distribution of Volunteers in the Geometrix Database by Age and Ancestry

Ancestry	Females	Males
White British	1265	1553
Other White background	56	78
White and Black Caribbean	4	3
White and Black African	1	2
White and Asian	4	6
Other Mixed background	6	3
Indian	14	17
Pakistani	4	11
Any other Asian background	6	11
Caribbean	2	8
African	7	7
Other Black background	3	1
Chinese	18	17
Any other	4	4
Total	1394	1721

FaceVision FV802 Series Biometric Camera, two XML files containing the repeated sets of landmark coordinates, and the 3D face surface data in three standard formats. The uncompressed database is 213GB in size.

The Cyberware database consists of 3D face surface and texture map data in Cyberware and TIFF format, respectively, for 1844 volunteers. All but 144

Table 4.7 Distribution of Volunteers in the Geometrix Database by Age Group and Sex

Age Group	Females	Males
14–19	194	188
20–24	71	68
25–29	98	103
30–34	173	188
35–39	291	325
40–44	247	361
45–49	127	181
50–54	53	111
55–59	55	65
60–64	44	54
65+	41	77
Total	1394	1721

of these volunteers were also scanned with the Geometrix FaceVision FV802 Series Biometric Camera. The uncompressed database is 2.2GB in size.

The 3D landmarks are stored in a Microsoft Office Excel 2003 spreadsheet, which holds the coordinate datasets for both repetitions of the landmark measurements for the 3,115 individuals in the Geometrix database. The uncompressed spreadsheet is 5.4MB in size.

4.4 Representativeness of the Database

In order to assess the general representativeness of the sample in the database, the effects of subgroups on the size and shape of faces was investigated.

4.4.1 Sex and Age

A generalized Procrustes analysis (see Dryden and Mardia 1998) was carried out in R (R 2008) on all the faces for which complete sets of landmarks were available. This process removed differences attributable to position and scaling, allowing shape and size to be analyzed via a principal components analysis (PCA). The first few PC scores are shown in Figures 4.2 to 4.5.

Figure 4.2 shows a plot of PC2 versus PC1 for males and females. There is a clear difference in PC1, which largely accounts for size variability. Also, it is noticeable that there is more overlap for the younger than older faces, and quite a big change in the male PC1 score in younger

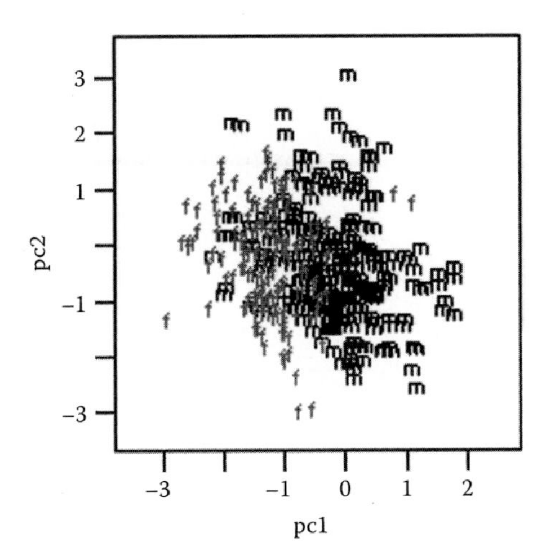

Figure 4.2 Plot of PC1 versus PC2 for males (black "m") and females (gray "f") for the under 25 age group.

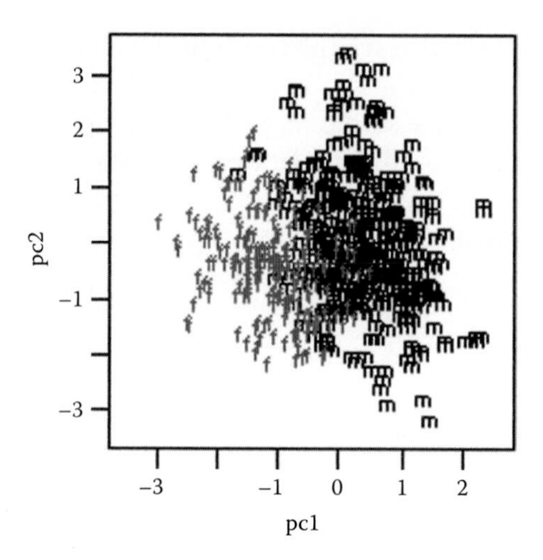

Figure 4.3 Plot of PC1 versus PC2 for males (black "m") and females (gray "f") for the 25 to 34 age group.

versus older faces, with the older male faces quite far from the females (see Figures 4.2 to 4.5). Hence, from these plots it is clear that male and female faces must be treated separately. There is a very clear gender effect, as one might expect.

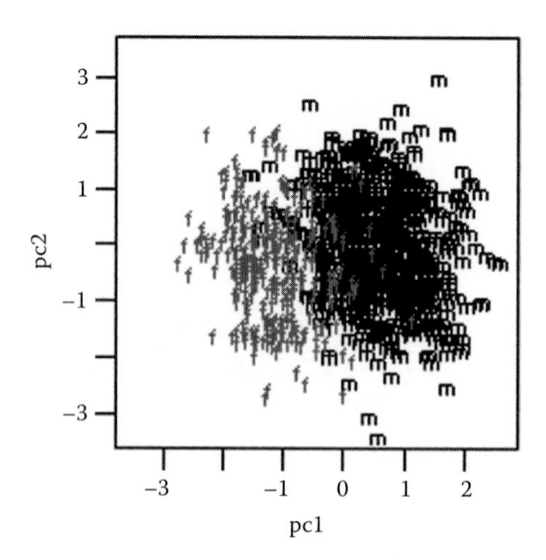

Figure 4.4 Plot of PC1 versus PC2 for males (black "m") and females (gray "f") for the 35 to 44 age group.

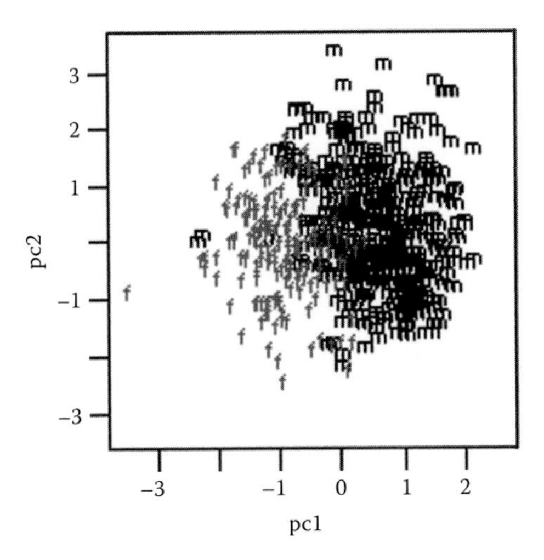

Figure 4.5 Plot of PC1 versus PC2 for males (black "m") and females (gray "f") for the 45 and over age group.

In order to investigate whether there is an important age effect, each sex must be considered separately in view of the above observations. Figures 4.6 to 4.9 show pair wise plots of the first few PC scores for males.

The most striking feature of these plots is that the youngest group has lower PC1 scores in general. PC1 can be interpreted as an overall size

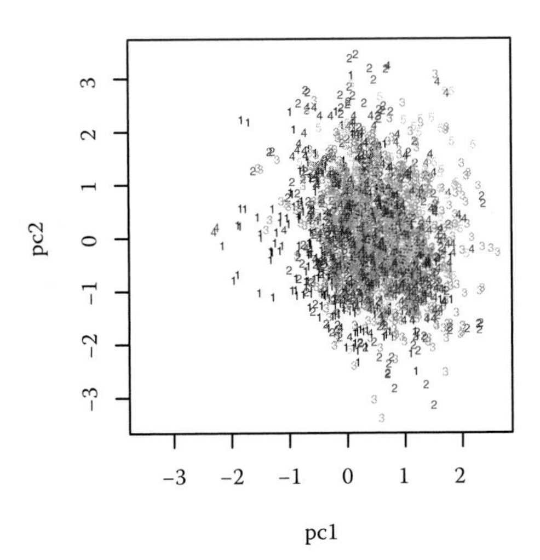

Figure 4.6 (See color insert following page 146.) Plot of PC1 versus PC2 for males by age group (1: under 25; 2: 25–34; 3: 35–44; 4: 45 and over).

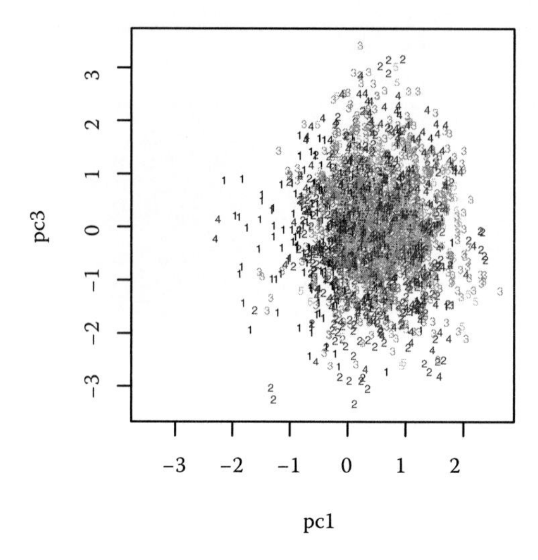

Figure 4.7 (See color insert following page 146.) Plot of PC1 versus PC3 for males by age group (1: under 25; 2: 25–34; 3: 35–44; 4: 45 and over).

component, accounting for 32.7% of the variability in the dataset of complete landmarks.

Figures 4.10 to 4.13 show pairwise plots of the first few PC scores for females, using the same age groups.

The difference between the younger and older female groups is not as clear as in males. There is a general tendency for PC1 scores to be higher for the older females, but this is not as dramatic as in the male case.

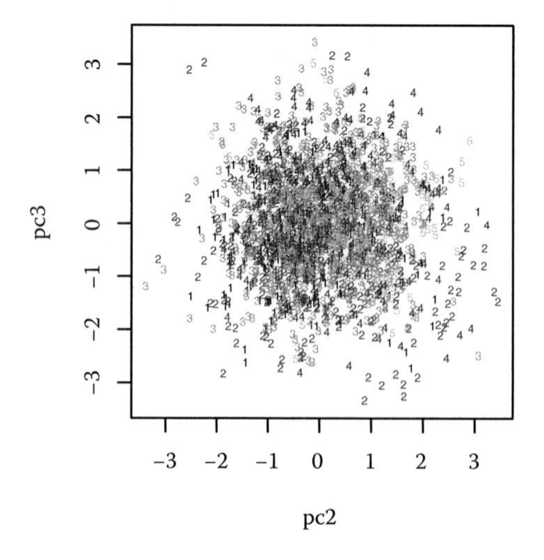

Figure 4.8 (See color insert following page 146.) Plot of PC2 versus PC3 for males by age group (1: under 25; 2: 25–34; 3: 35–44; 4: 45 and over).

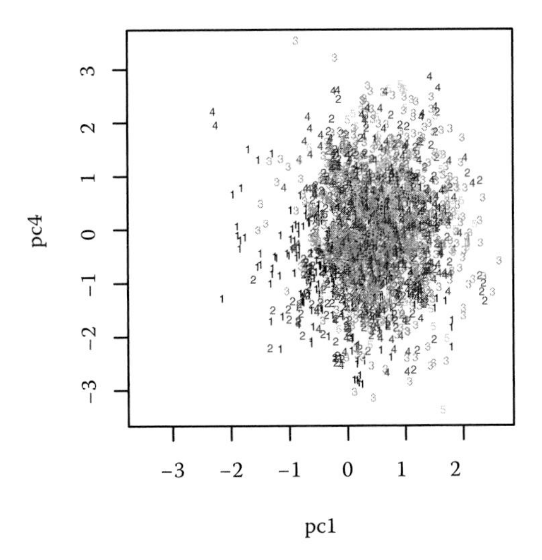

Figure 4.9 (See color insert following page 146.) Plot of PC1 versus PC4 for males by age group (1: under 25; 2: 25–34; 3: 35–44; 4: 45 and over).

4.4.2 Ancestry

To investigate whether ancestry leads to different sizes and shapes of faces, a further PCA was carried out. Figure 4.14 to 4.17 show plots of the first two PC scores for the U.K. Census categories White, Mixed, Asian, Black, and, arbitrarily, Other.

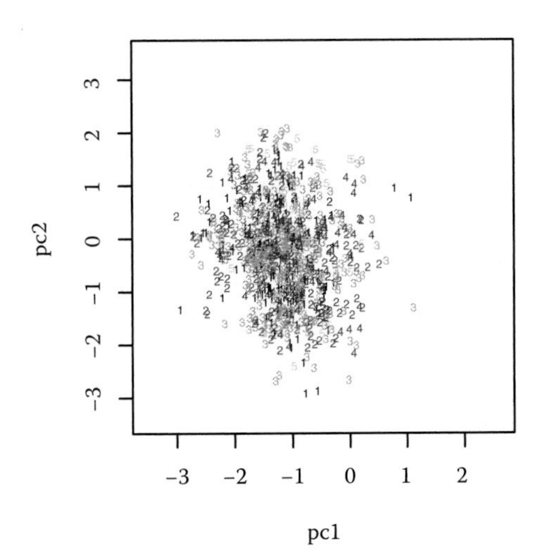

Figure 4.10 (See color insert following page 146.) Plot of PC1 versus PC2 for females by age group (1: under 25; 2: 25–34; 3: 35–44; 4: 45 and over).

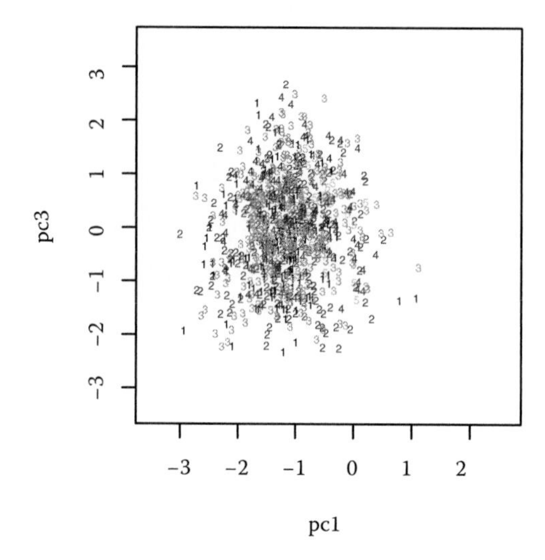

Figure 4.11 (See color insert following page 146.) Plot of PC1 versus PC3 for females by age group (1: under 25; 2: 25–34; 3: 35–44; 4: 45 and over).

There is no clear pattern to these values in the first few PCs, indicating that ancestry is not an overall factor in size and shape variation when all of the available landmarks are considered in total. The sample is overwhelmingly from the U.K. Census category White British, however, and other ancestries are poorly represented. Furthermore, when taken separately, some subsets of landmarks may show ancestry related patterns.

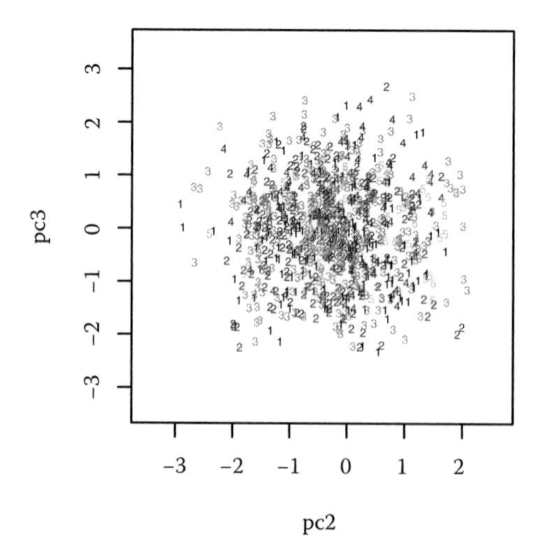

Figure 4.12 (See color insert following page 146.) Plot of PC2 versus PC3 for females by age group (1: under 25; 2: 25–34; 3: 35–44; 4: 45 and over).

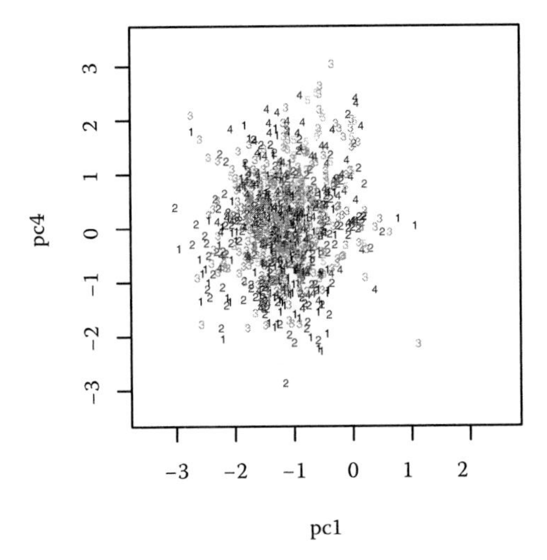

Figure 4.13 (See color insert following page 146.) Plot of PC1 versus PC4 for females by age group (1: under 25; 2: 25–34; 3: 35–44; 4: 45 and over).

4.5 Frequencies

As a very simple approximation to some form of "face frequency" measure a parametric distribution was fitted to the face data separately for each sex and age group. On the basis of the landmark measurements for the individual, the following measure of face frequency was considered:

Let *p_i = P(sum of first p squared PC scores for the ith face|chi-squared_p distribution).*

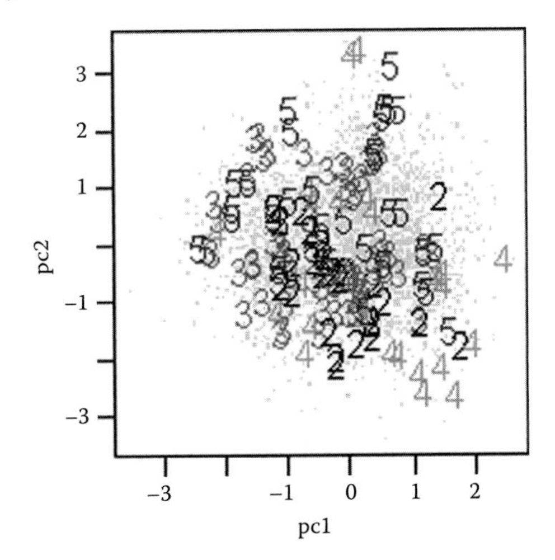

Figure 4.14 (See color insert following page 146.) Plot of PC1 versus PC2 for the U.K. Census categories: White (light dots), Mixed (black "2"s), Asian (gray "3"s), Black (light gray "4"s), Other (gray "5"s).

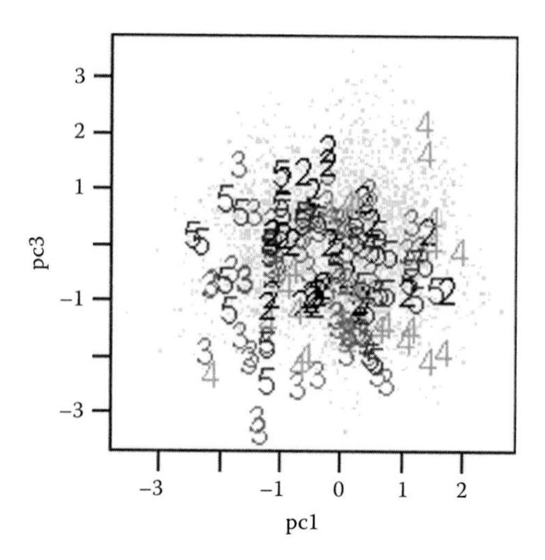

Figure 4.15 (See color insert following page 146.) Plot of PC1 versus PC3 for the U.K. Census categories: White (light dots), Mixed (black "2"s), Asian (gray "3"s), Black (light gray "4"s), Other (gray "5"s).

The estimated probability of the *i*th face (face frequency) is $f_i = p_i/sum(p_i)$.
It is often easier to interpret this number as a frequency of "1 *in* N_i," where $N_i = 1/f_i$.

Figure 4.18 shows a histogram of the sum of the first 20 squared PC scores (standardized PC scores) for males aged 15 to 24, which under a multivariate

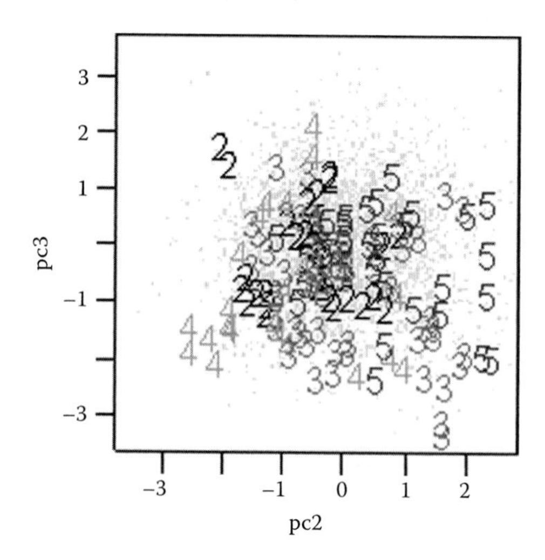

Figure 4.16 (See color insert following page 146.) Plot of PC2 versus PC3 for the U.K. Census categories: White (light dots), Mixed (black "2"s), Asian (gray "3"s), Black (light gray "4"s), Other (gray "5"s).

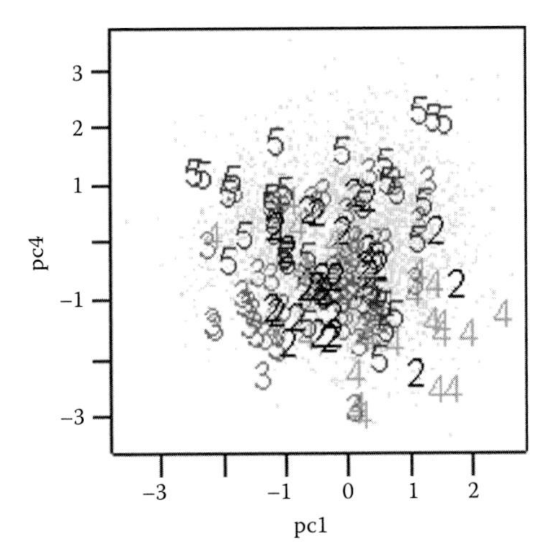

Figure 4.17 (See color insert following page 146.) Plot of PC1 versus PC4 for the U.K. Census categories: White (light dots), Mixed (black "2"s), Asian (gray "3"s), Black (light gray "4"s), Other (gray "5"s).

normal model should have a chi-squared distribution with 20 degrees of freedom (df; marked as a gray line in the figure).

Twenty df was chosen as a high proportion of variability was explained by the first 20 PCs. The theoretical distribution seems reasonable and so this is used to calculate the face frequency. For example, for males aged 15 to 24 the following range of estimated face frequencies may be calculated:

Individual "190405_00026" landmark set 1 $N\_i = 17332$
Individual "190405_00026" landmark set 2 $N\_i = 10380$

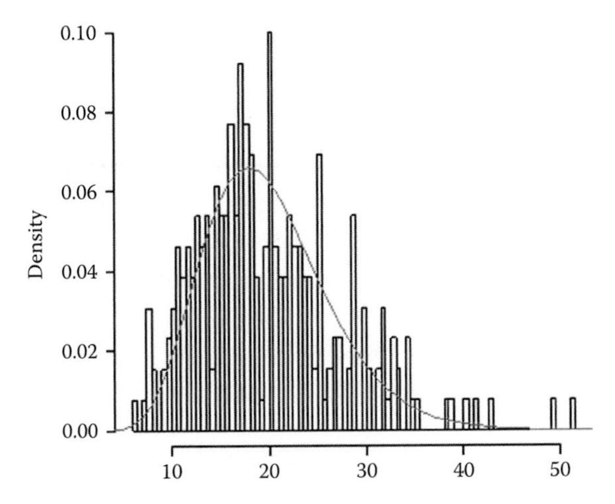

Figure 4.18 Histogram of the sum of the first 20 PC scores for males aged 15–24. The gray line shows the fitted chi-squared df = 20 distribution used to calculate the face frequencies.

Individual "291004_00011" landmark set 1 $N\_i = 970$
Individual "291004_00011" landmark set 2 $N\_i = 884$
Individual "300804_00013" landmark set 1 $N\_i = 447$
Individual "300804_00013" landmark set 2 $N\_i = 356$
Individual "190405_00030" landmark set 1 $N\_i = 172$
Individual "190405_00030" landmark set 2 $N\_i = 223$

Face frequency is quoted as *1 in N_i*, with larger *N_i* meaning a rarer face. So, in these examples the first face is the rarest by far, then the second, and then the third, and finally the fourth face is the most common.

The differences between the values of *N_i* for the two independent landmark datasets of the same face are interesting and offer an indication of the influence of observer error on the estimation of frequency.

Note that the value of *N_i* should be interpreted with care. It is a crude measure, and very prone to measurement and model misspecification error, particularly for unusual rare faces. However, with care it should provide a useful order of magnitude type measure of face frequency—as comparison of the two independent landmark datasets from each face appears also to indicate.

4.6 Summary

A large sample database of over 3000 volunteers' 3D facial images was collected, and each facial image was landmarked in duplicate at up to 30 landmark sites.

The volunteers were predominantly White British according to the U.K. Census categories, although many other categories were represented. The U.K. Census categories are self-declared, and provide a loose approximation with ancestry. Males predominate in the database, as do people in middle age, meaning in this case 35 to 44 years of age, although younger age groups are reasonably well represented. The distribution is probably a reflection of the profile of typical visitors to a popular science museum in South Yorkshire, England.

Face shape and size are clearly affected by sex. The affect of age is greater in males. Size appears to be a significant factor affecting sex and age based variation.

There is some evidence that ancestry, in this case represented by U.K. Census category, influences face shape, but the number of non-White British volunteers in the sample was small.

A crude estimate of face shape frequencies indicates that a huge range of frequencies might be anticipated, in the range of one in a few hundred to one in a few tens of thousands.

References

Dryden, I. L., and K. V. Mardia, 1998. *Statistical shape analysis*. London: Wiley.
Hartman, J., and J. Wernecke, 1996. *The VRML 2.0 handbook: building moving worlds on the Web*. New York: Addison Wesley.
R., 2008. *The R project for statistical computing*. http://www.r-project.org (accessed June 30, 2009).

Investigation of Anthropometric Landmarking in 2D

5

LUCY MORECROFT, NICK R.J. FIELLER, AND MARTIN PAUL EVISON

Contents

5.1 Introduction

Until public area or security cameras are able to routinely and reliably capture 3D images, facial image comparison will rely on 2D images, whether or not they are to be compared with a 2D or 3D alternative.

Tools for anthropometric facial comparison that use these images will inevitably rely on the manual or automatic identification of features or landmarks in them. A computer-assisted method of analysis will be likely to employ manual anthropometric landmark placement using a software tool. Image quality can be predicted to have a strong influence on the accuracy of anthropometric landmark placement. Poor image quality will result in landmarks that are inaccurately or imprecisely placed or cannot be placed at all, in turn influencing the value of any empirical comparison between faces.

In this chapter, an investigation of anthropometric landmarking of facial images in 2D is described. The investigation was directed at establishing whether individuals can be distinguished using anthropometric measurements derived from a range of 2D anterior photographs of the same subject, irrespective of the influence of observer error, and to provisionally assess to what extent image quality affects the accurate placement of landmarks in 2D.

This chapter also describes a simple experiment designed to provide a provisional assessment of image quality on landmark visibility in 2D in order to establish guidelines from which standards might be derived.

5.2 Facial Image Samples Used

Two 2D facial image samples were prepared, a primary sample to assess utility in distinguishing individual subjects and the influence of error, and the secondary sample to examine the affects of image quality. The camera used was a Nikon Coolpix® 5700.

The primary sample consisted of anterior 2D facial images of five different individuals (subjects A, B, C, D, and H). Each subject was photographed three times (photographs 0, 1, and 2) at different intervals over the period of one

Figure 5.1 Location of the 30 anthropometric landmarks used in the investigation shown in an inferior anterior-lateral or three-quarter profile view. Bilateral landmarks are marked with an asterisk.

Table 5.1 List of the Anthropometric Landmarks Used in the Investigation and Shown in Figure 5.1

Label	Name	Description (see Farkas 1994)
g	Glabella	The most prominent midline point between the eyebrows
sl	Sublabiale	Determines the lower border of the lower lip and upper border of the chin
pg	Pogonion	The most anterior midpoint of the chin
en	Endocanthion (l, r)	The point at the inner commissure of the eye fissure
ex	Exocanthion (l, r)	The point at the outer commissure of the eye fissure
p	Pupil (l, r)	Determined when the head is in the rest position and the eye is looking straight forward
pi	Palpebrale inferius (l, r)	The lowest point in the mid portion of the free margin of each lower eyelid
se	Sellion	The deepest landmark located in the bottom of the nasofrontal angle
prn	Pronasale	The most protruded point of the apex nasi
al	Alare (l, r)	The most lateral point on each alar contour
c′	Highest point of columella prime (l, r)	The point on each columella crest, level with the tip of the corresponding nostril
ls	Labiale superius	The midpoint of the upper vermillion line
li	Labiale inferius	The midpoint of the lower vermillion line
sto	Stomion	The imaginary point at the crossing of the vertical facial midline and the horizontal labial fissure between gently closed lips, with the teeth shut in the natural position
ch	Cheilion (l, r)	The point located at each labial commissure
sa	Superaurale (l, r)	The highest point on the free margin of the auricle
sba	Subaurale (l, r)	The lowest point on the free margin of the ear lobe
pa	Postaurale (l, r)	The most posterior point on the free margin of the ear
obi	Otobasion inferius (l, r)	The point of attachment of the ear lobe to the cheek

day, giving a total of 15 images. Three different landmark placement operators (observers black, light gray, and dark gray) each placed the standard 30 anthropometric landmarks (Figure 5.1 and Table 5.1) on each of the 15 images using the 2D landmarking tool (see Chapter 11); they repeated this process three times for each photo (repetitions 0, 1, and 2), giving a total of 45 configurations per observer and a grand total of 135 facial image landmark datasets for analysis.

The secondary sample consisted of a similar set of anterior 2D facial images of five different individuals. Each face was photographed once and eleven other copies of this baseline image were made. The quality of the eleven copied images was reduced by diminishing one or more of four factors: resolution, focus, compression, and illumination. This produced 12 images of varying quality for each of the five subjects under investigation with a total of 60 images. For each factor, the extent of diminished quality used in the analysis was somewhat arbitrarily chosen, but represented a clear change in appearance of the image, neither so trivial that it was not visible nor so extreme that facial features were clearly distorted or obliterated. In each case, the measure was quantified.

The 2D coordinates of the standard 30 facial landmarks (Figure 5.1) were again collected. Three landmark placement operators each placed 30 landmark points on each of the 60 images. They each repeated this process three times, giving a total of 540 landmark configurations for analysis.

5.3 Parameters of Image Quality

5.3.1 Resolution

The greater the number of pixels used to represent an image, the greater will be its fidelity with the object it represents; the fewer pixels used, the poorer the fidelity will be. With lack of fidelity there will be concomitant reduction in the accuracy and precision of anthropometric landmark placement.

The number of pixels determines the resolution of the image. The resolution of the region of interest (the face) may differ from that of the whole image, depending on its location. For example, if there are two faces in an image, one close, one far away, then the closer face will have a higher resolution than the one farther away.

The measure of resolution chosen in this analysis was the area (in pixels^2) of a rectangle enclosing the head as illustrated in Figures 5.2 and 5.3. For comparison, the diminished quality image was adjusted to approximately

Figure 5.2 Image at "high" resolution: 40,179 pixels^2.

Figure 5.3 Image at "low" resolution: 2508 pixels^2.

25,000 pixels^2 and the unadjusted image was used as the normal control. Other measures proposed, such as the number of pixels between the centers of the eyes, are dependent on the pose angle of the subject and could not be employed in this dataset.

5.3.2 Focus

Measurement of focus relied on methods used in the "autofocus" feature of many cameras. Autofocus algorithms work by maximizing a measure of the localized contrast in the image. When the image is soft due to over- and underfocusing, the pixel values vary little in a small neighborhood and the contrast value will be low; when the image is in focus, the contrast value will be at its maximum.

A simple contrast measure, C_{h+v} (see Equation 5.1), was adopted, which is a combination of both horizontal and vertical contrast values (cameras often use only contrast between neighboring horizontal pixels):

$$C_{h+v} = \frac{1000}{MN(f^{hi} - f^{lo})}\left(\sum_{i=0,j=0}^{M-1,N-2} |f_{i,j} - f_{i,j+1}| + \sum_{i=0,j=0}^{M-2,N-1} |f_{i,j} - f_{i+1,j}| \right) \tag{5.1}$$

Where f is the image of size $M \times N$, f^{lo} is the gray level at which ~12.5% of pixels have intensity $< f^{lo}$, and f^{hi} is the gray level at which ~12.5% of pixels have intensity $> f^{hi}$. The values f^{lo} and f^{hi} are included in an attempt to account for

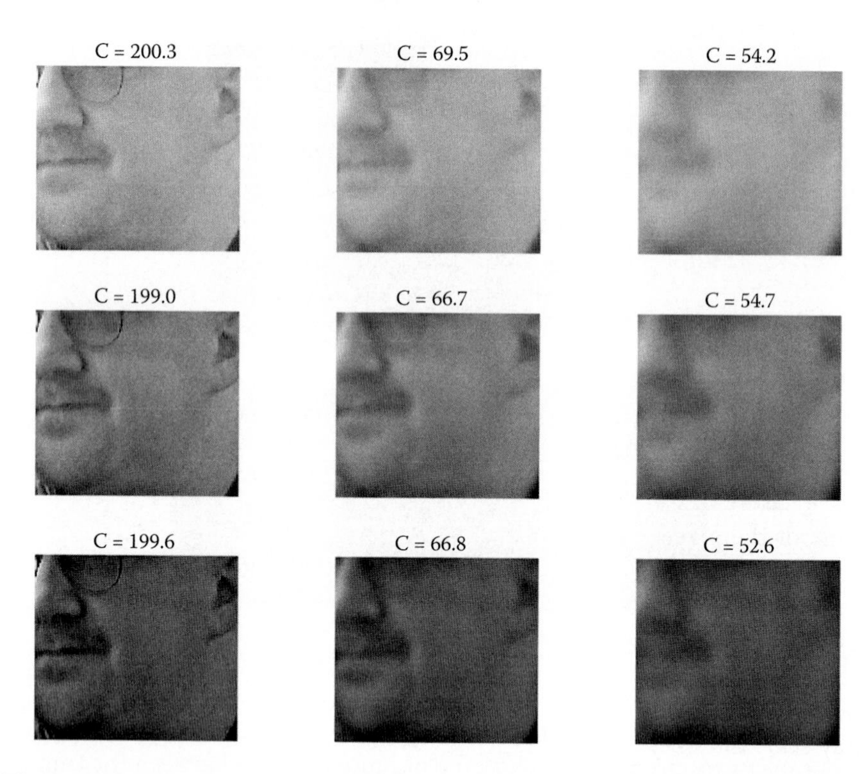

Figure 5.4 Measuring focus under varying lighting conditions. The top row of images has been brightened using a gamma correction factor of 2.0, the bottom row has been darkened with a gamma correction factor of 0.5 and the middle row has not had its brightness adjusted. The left column has not been defocused, the middle column has been "defocused" slightly (convolved with Gaussian of radius 2.0), and the right column has been "defocused" more (convolved with Gaussian of radius 4.0). The resulting values of *C* are fairly robust to changes in lighting.

the overall lighting conditions, which could otherwise influence the contrast measurements.

Care was taken not to include any overly dark or saturated areas as these can skew the results. An area of the face with a good amount of detail was selected. The area chosen only included the subject's face or head. Background and other objects were excluded as their level of focus will be different (see Figure 5.4).

In the analysis, an unadjusted control with no blurring was used in comparison with a diminished quality image obtained using a convolution with a Gaussian of width 7.

5.3.3 Compression

Digital image compression is complex, with a variety of processes available including lossless and a variety of forms of lossy compression. Without an original uncompressed image for reference, the usual compression distortion measures such as peak signal-to-noise ratio (PSNR) and mean square error (MSE) are impossible to obtain. In such cases, the best available measures to quantify the degradation due to compression are those which indicate the amount of compression the image has undergone. To provide an approximate parameter for compression, Bits per pixel (BPP) may be calculated from the compressed image file size and pixel dimensions. In this analysis, however, the influence of compression was examined using an image compressed in JPEG format to image quality 90 (out of 100) to provide a diminished quality image. The uncompressed image was used as a comparative control.

5.3.4 Illumination

The measure of illumination or brightness chosen was the mean gray scale value of the pixels contained in the largest rectangle that could be placed within the head (see Figure 5.5), taking care to avoid any obviously saturated areas. Choosing the rectangle inside the head ensured that the subject's background did not affect the brightness measurement. Grayscale (8-bit) values were used with black = 0 and white = 255. In this analysis, the photograph's own natural brightness level was compared with an image set to an average gray scale intensity of 50, obtained by subtracting an appropriate constant value from each pixel.

Figure 5.5 Measuring the brightness of the subject of the image. The area selected was the largest rectangle that could be enclosed by the subject's face, taking care to avoid obviously saturated areas. Mean intensity values in the selected areas are 144 (left) and 88 (right).

5.4 Analysis of Face Shape Variation and Observer Error

The method of comparison between landmark datasets utilizes Procrustes registration, which requires there be no missing values in the data. An initial examination of the primary facial image sample revealed a number of missing values, however. None of the observers had placed the bilateral landmarks of the ear, as locating them is impossible in the 2D anterior view (for illustration, compare Figures 5.1 and 5.9). One operator had not placed the highest point of columella prime left and right on a number of images, and so these landmarks were discarded in the entire sample. Procrustes registration was completed without scaling to enable direct comparison between the landmark datasets via a principal components analysis (PCA) of the Procrustes residuals in R (R 2008).

Outliers were detected in a preliminary PCA (Figure 5.6) and were found to be due to two random observer errors. In the first, bilateral landmarks of

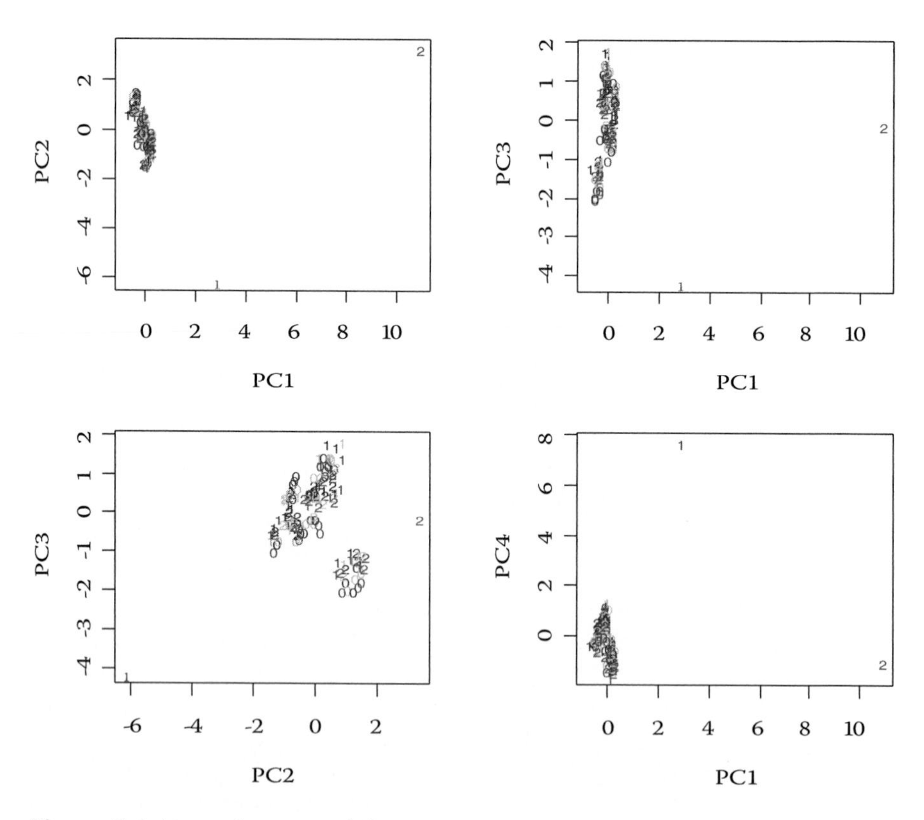

Figure 5.6 (See color insert following page 146.) Results of a preliminary PCA analysis of the primary 2D facial image sample showing the first four PCs, permitting comparison of landmark datasets by subject (subjects A, B, C, D, and H) and observer (represented by three grayscale values in the figure). Some outliers are visible as black "C"s in the figure representing observer black and subject C.

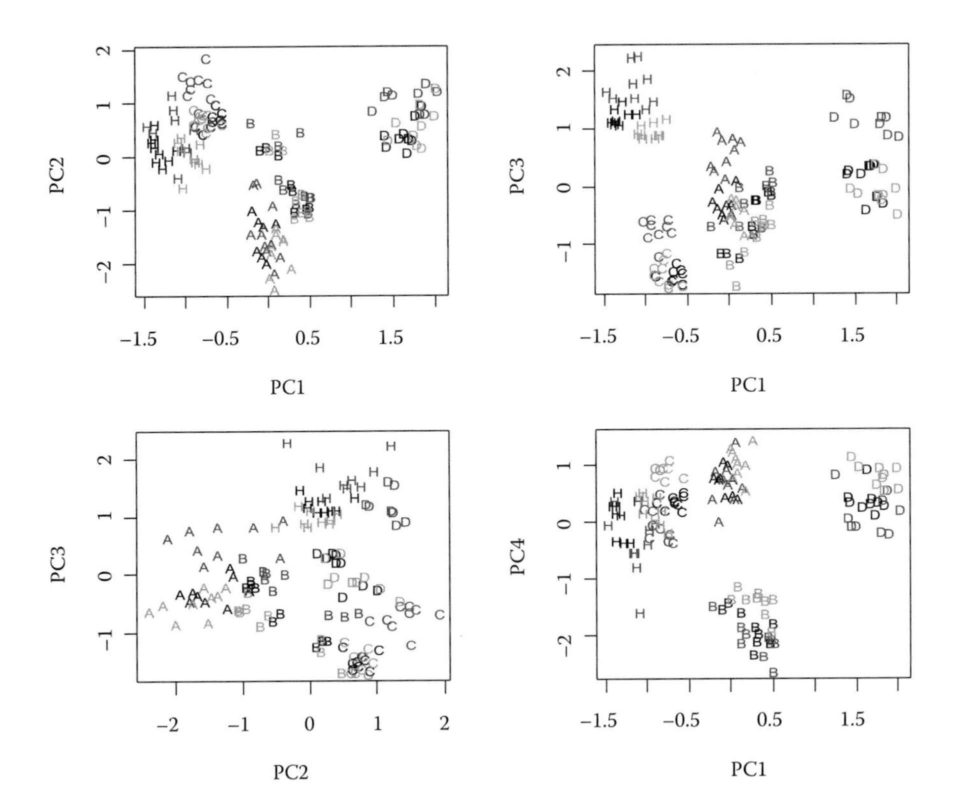

Figure 5.7 (See color insert following page 146.) Results of a PCA analysis on data corrected for outliers of the primary 2D facial image sample showing the first four PCs, which account for 76.4% of variation. This analysis permits comparison of landmark datasets by subject (subjects A, B, C, D, and H) and observer (represented by three grayscale values in the figure).

the exocanthion (see Figure 5.1) had been inadvertently placed on the wrong side; an error easily rectified by swapping over the left and right coordinates. In the second, an *x*-coordinate was found to be very different to that of the other two repetition measurements taken at the same landmark. An average of these latter values was used to rectify this error. As a consequence, variation at these landmarks for this observer is likely to be underestimated.

PCA of the sample corrected for outliers (Figure 5.7) indicates that the landmark datasets for subjects A, B, C, D, and H cluster together even with repeated landmarkings by different observers of different photographs of the subject.

Careful scrutiny of Figure 5.7 indicates that landmark datasets for each subject cluster more closely for some observers than for others.

A further PCA (Figure 5.8) was used to assess variation due to differences between repeated landmarkings of different photographs of the same subject.

Figure 5.8 shows little evidence of clustering due to photograph number or repetition number. The random dispersion implies that the two variables have no

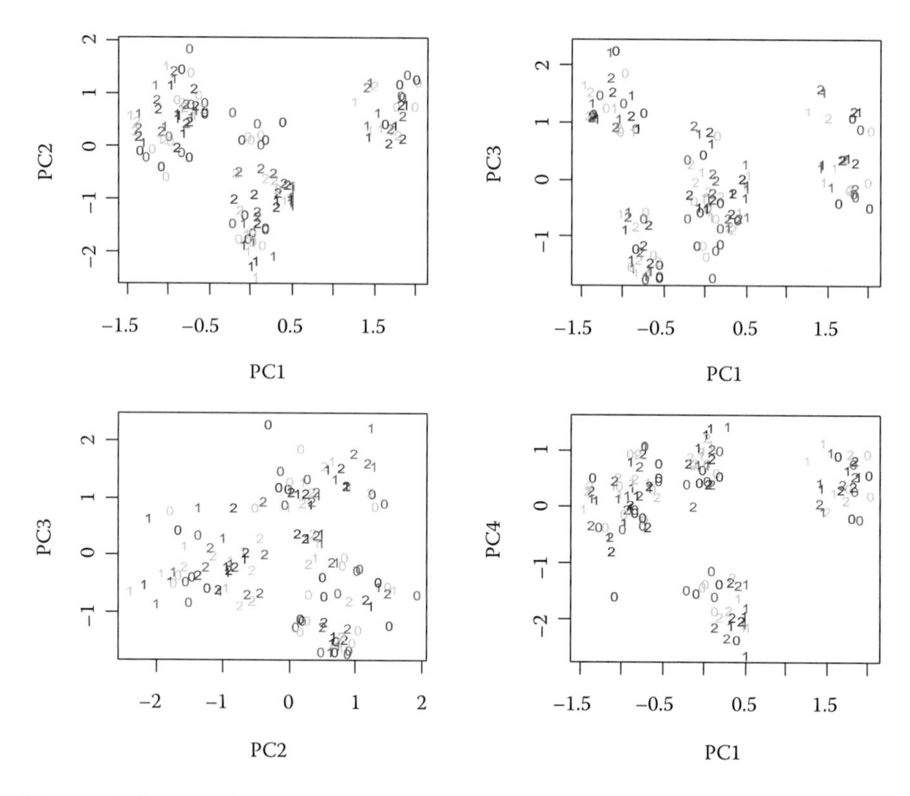

Figure 5.8 (See color insert following page 146.) Results of a PCA analysis on data corrected for outliers of the primary 2D sample showing the first four PCs. This analysis permits comparison of landmark datasets by photograph of the same subject (photographs 0, 1, and 2) and landmarking repetitions (represented by three grayscale values in the figure).

effect on the resulting landmark configurations. Careful scrutiny of Figures 5.7 and 5.8 reveals, for subject B, repetitions of photograph 0 cluster separately from those of photographs 1 and 2, which are more dispersed. This is particularly clear in the first PC, which accounts for about 55% of the variation.

The three photographs of subject B are shown in Figure 5.9. Some variation in pose angle and facial expression is evident, which might be hypothesized

Figure 5.9 Three photographs (0, 1, and 2) of subject B. Note the differences in pose angle and facial expression.

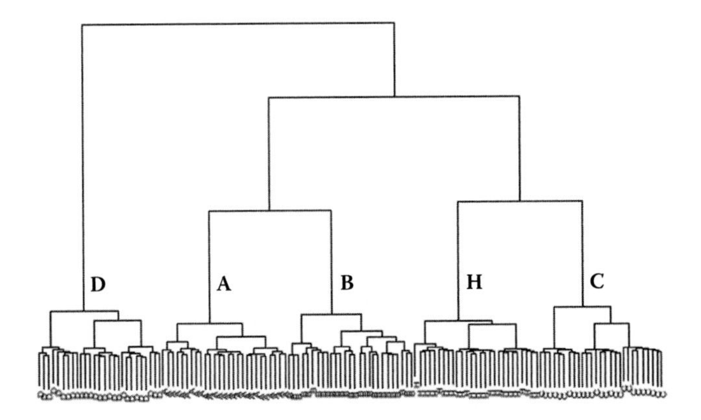

Figure 5.10 Dendrogram showing clustering of facial image landmark datasets by subject. For clarity, branches are labeled with the subject identifier.

as elevating differences between the landmark datasets of photograph 0, and photographs 1 and 2.

In order to assess whether facial image landmark datasets from the same subject grouped together and independently from those of other subjects, a cluster analysis was undertaken. Ward's method was used on the Procrustes-registered data corrected for outliers.

The primary clustering is by subject (see Figure 5.10).

Figures 5.11 and 5.12 show dendrograms indicating clustering of facial image landmark datasets by photograph and repetition number, respectively.

Figures 5.11 and 5.12 indicate that datasets representing repeated landmarkings of the same photograph tend to cluster together, but close scrutiny shows that this process is by no means uniform. For every subject, there are

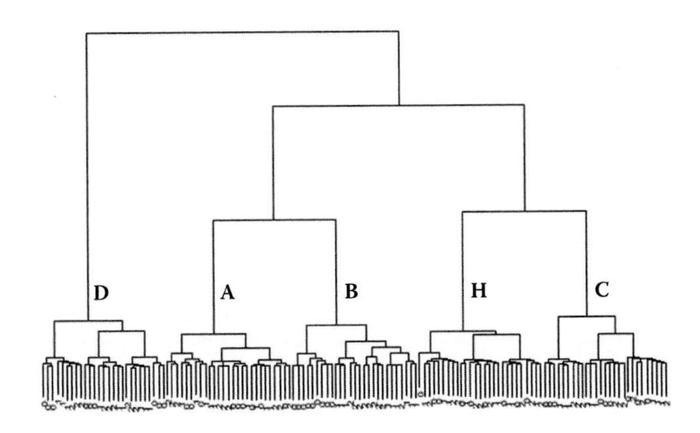

Figure 5.11 Dendrogram showing clustering of facial image landmark datasets by photograph number.

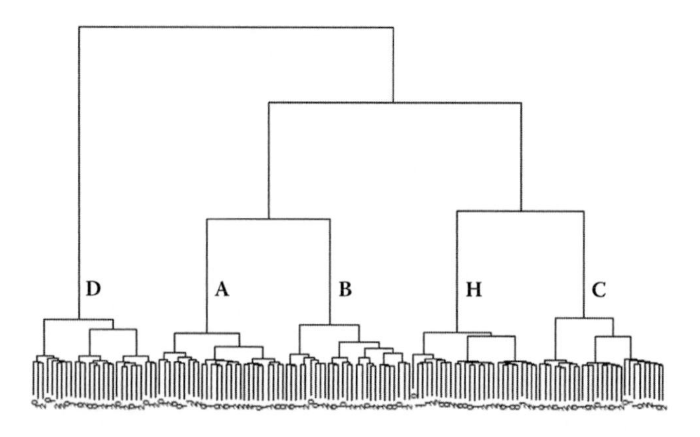

Figure 5.12 Dendrogram showing clustering of facial image landmark datasets by repetition number.

examples where an occasional repetition is found to cluster more closely to those of another photograph of that subject. Landmark datasets always cluster within subject, however.

Figure 5.13 shows a dendrogram indicating clustering by observer.

Figure 5.13 shows that landmark datasets tend to cluster within subject by observer. Again, close scrutiny indicates this clustering is not entirely uniform. For subject B, observer landmark datasets do not cluster together. Comparison with Figure 5.11 indicates that in this case it is the landmark datasets of photograph 0 of subject B that are clustering together, and apart

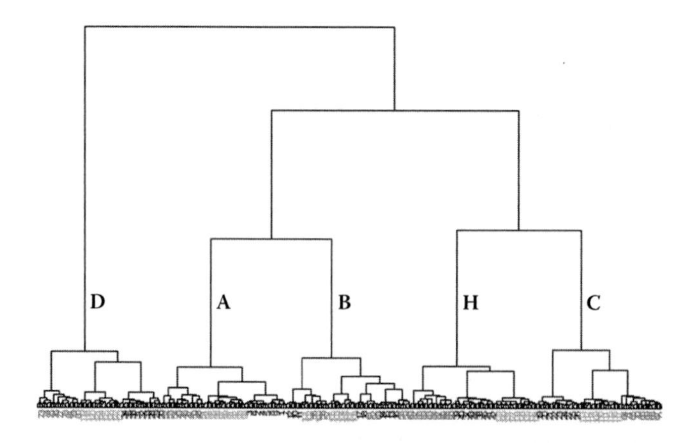

Figure 5.13 Dendrogram showing clustering of facial image landmark datasets by observer (each dataset is represented by a dataset number shown in gray scale according to observer).

from those of photographs 1 and 2. This observation is also evident in the PCA analysis (see Figure 5.8).

Photograph 0 of subject B (see Figure 5.9) appears to be sufficiently distinctive relative to the other two photographs, and relative to variation in the photographs of the other subjects, that it clusters independently irrespective of observer or repetition. Photographs of the other subjects appear to be sufficiently similar to permit variation in operator practice to cause facial image landmark datasets to cluster by observer. This variation or error may be systematic. Nevertheless it is sufficiently subtle not to affect a uniform and fundamental clustering that occurs by subject: irrespective of photograph, repetition, and observer, the landmark datasets of the five subjects always cluster together. All subjects can be differentiated one from another in all datasets, irrespective of observer.

5.5 Analysis of Image Quality and Landmark Placement

A large number of missing values in the secondary dataset precluded an analysis of accuracy or error in landmark placement comparable to that used in above. The fact that the landmarks are missing is of interest in itself, however, and permits the influence of image quality degradation on landmark placement to be assessed.

Table 5.2 shows the percentage of missing values for each image quality test, along with the levels (normal or degraded) of resolution, focus, illumination, and compression used in each test.

Table 5.2 Percentage of Missing Data for Each Image Quality Test

Test	Resolution	Focus	Compression	Illumination	Percent of Data Missing
0	Unadjusted	Unadjusted	Unadjusted	Unadjusted	9
1	Unadjusted	Unadjusted	JPEG Q = 90	Unadjusted	10
2	Unadjusted	Unadjusted	Unadjusted	Average grayscale = 50	10
3	Unadjusted	Unadjusted	JPEG Q = 90	Average grayscale = 50	15
4	Unadjusted	Gaussian 7	Unadjusted	Unadjusted	9
5	Unadjusted	Gaussian 7	JPEG Q = 90	Unadjusted	11
6	Unadjusted	Gaussian 7	Unadjusted	Average grayscale = 50	11
7	Unadjusted	Gaussian 7	JPEG Q = 90	Average grayscale = 50	15
8	~25K pixels^2	Unadjusted	Unadjusted	Unadjusted	10
9	~25K pixels^2	Unadjusted	JPEG Q = 90	Unadjusted	19
10	~25K pixels^2	Unadjusted	Unadjusted	Average grayscale = 50	11
11	~25K pixels^2	Unadjusted	JPEG Q = 90	Average grayscale = 50	29

Test 0, using unaltered control images, has the least missing data at 9%. Test 4, which has an adjustment for focus and none for the other three quality factors, shows the same amount of missing data, however. This implies that degrading the focus of an image alone probably does not affect the visibility and placement of these landmarks in an anterior 2D image.

Degrading the other quality factors one by one (compression in Test 1, illumination in Test 2, resolution in Test 8) each resulted in a 1% increase of missing data when compared to the control.

Tests involving two adjustments and two unchanged factors (Tests 3, 5, 6, and 10) resulted in varying percentages of missing data. Tests 5 and 6 show that adjusting focus in the image in addition to either compression or illumination, respectively, only results in a 1% increase of missing landmarks, as does Test 10, when resolution is adjusted in addition to illumination. Test 3 shows that a combination of degradation of illumination and compression results in a 6% increase of missing data when compared to the control.

By far the worst quality test in terms of incomplete data is Test 11 with 29% missing data. This is to be expected as it has three out of the four quality factors set at adjusted levels: resolution, illumination and compression. Interestingly, Test 7 also has three extreme adjustments: focus, illumination and compression; but only has 15% of the data incomplete. Poor image resolution has a worse affect on landmark visibility and placement than poor focus.

It is useful to know which of the 30 landmarks are easily located in anterior 2D images and which are not. Table 5.3 summarizes the percentage of missing data in each quality test by landmark, rounded to the nearest whole number.

There was no data missing for landmarks 1 to 13 (see Table 5.3) in all tests except Test 11. These landmarks are points on the facial midline and around the eye area. Despite the degraded image quality, therefore, these points were thought visible and placed by the operators.

Landmarks 14 to 22 were visible in most tests, apart from Tests 3, 7, 9, and 11. These landmarks are located around the nose and mouth. All the tests that had missing data for these landmarks had in common an adjustment for compression. The compression of an image, when combined with other degrading factors, appears to have an effect on the visibility of facial landmarks around the nose and mouth, but not on those around the facial midline and the eyes.

Landmarks 23 to 30 have missing values in all tests. These are landmarks around the ears and are not clearly visible in the 2D anterior view per se.

Table 5.3 Percentage of Missing Data for Each Image Quality Test by Landmark

	Test												
Landmark	0	1	2	3	4	5	6	7	8	9	10	11	All
1 Glabella	0	0	0	0	0	0	0	0	0	0	0	2	0
2 Sublabiale	0	0	0	0	0	0	0	0	0	0	0	27	2
3 Pogonion	0	0	0	0	0	0	0	0	0	0	0	27	2
4 Endocanthion l	0	0	0	0	0	0	0	0	0	0	0	11	1
5 Endocanthion r	0	0	0	0	0	0	0	0	0	0	0	9	1
6 Exocanthion l	0	0	0	0	0	0	0	0	0	0	0	11	1
7 Exocanthion r	0	0	0	0	0	0	0	0	0	0	0	11	1
8 Pupil l	0	0	0	0	0	0	0	0	0	0	0	11	1
9 Pupil r	0	0	0	0	0	0	0	0	0	0	0	11	1
10 Palpebrale inferius l	0	0	0	0	0	0	0	0	0	0	0	11	1
11 Palpebrale inferius r	0	0	0	0	0	0	0	0	0	0	0	16	1
12 Sellion	0	0	0	0	0	0	0	0	0	0	0	16	1
13 Pronasale	0	0	0	0	0	0	0	0	0	0	0	18	1
14 Alare l	0	0	0	7	0	0	0	4	0	9	0	31	4
15 Alare r	0	0	0	4	0	0	0	4	0	11	0	31	4
16 Highest point of columella prime l	0	0	0	11	0	0	0	11	0	18	0	33	6
17 Highest point of columella prime r	0	0	0	24	0	0	0	24	0	24	0	33	9
18 Labiale superius	0	0	0	4	0	0	0	4	0	7	0	27	4
19 Labiale inferius	0	0	0	4	0	0	0	7	0	7	0	27	4
20 Stomion	0	0	0	4	0	0	0	7	0	7	0	27	4
21 Cheilion l	0	0	0	4	0	0	0	4	0	9	0	27	4
22 Cheilion r	0	0	0	4	0	0	0	4	0	16	0	27	4
23 Superaurale l	40	40	40	44	40	40	40	42	40	49	40	56	43
24 Superaurale r	38	40	38	40	38	38	38	38	38	49	38	58	41
25 Subaurale l	7	16	18	27	9	18	20	22	20	22	22	38	20
26 Subaurale r	16	18	18	20	22	18	20	18	22	16	22	29	20
27 Postaurale l	33	36	47	44	29	40	47	47	38	47	49	60	43
28 Postaurale r	31	38	47	44	36	42	49	47	40	47	49	58	44
29 Otobasion inferius l	51	53	51	62	51	53	51	67	53	67	51	67	56
30 Otobasion inferius r	51	53	51	62	51	53	51	67	53	67	51	67	56

5.6 Summary of Findings

5.6.1 Face Shape Variation and Observer Error

Care in landmark placement is essential as gross errors—placing bilateral landmarks on the wrong side, for example—have a major influence on variation.

Repeated landmarking by different observers is essential to permit detection of outliers and related random errors, and to perform quality assurance checks on landmarking practice within and between operators.

Some variation in practice or systematic observer error in landmark placement is evident in this small study. Standard operating procedures and thorough training is desirable to minimize or eliminate this variation, although it did not prevent individual subject's faces from being uniformly distinguished by cluster analysis in this small study.

When random errors identified as outliers were rectified, facial image landmark datasets from different photographs of the same subject clustered persistently together, irrespective of photograph, observer, or repetition. This suggests that comparison by anthropometric landmarks offers a promising method of distinguishing faces of individuals.

5.6.2 Image Quality and Landmark Placement

A facial image that is highly out of focus, highly pixelated due to low resolution, distorted or obliterated by compression, or obscured by lack of illumination or blown out by excessive illumination will very probably be useless for anthropometric comparison. This investigation examined more marginal, if visible, changes.

Our findings suggest, paradoxically, that partially out of focus images may be adequate for anthropometric landmark-based comparison, even when either resolution or illumination are also diminished, or the image is compressed with loss. In contrast to diminished focus, compression compounds the effects of the other two factors, resolution and illumination, both independently and, most noticeably, in combination.

If reduction in image quality can prevent placement of landmarks by making them invisible to the operator, partial reduction may result in placement which is inaccurate or imprecise. Although this analysis did not address this issue directly, the above observations are also likely to apply to accuracy and precision of landmark placement generally.

These general implications may be complicated when individual landmarks are considered. Craniofacial landmarks are 3D features that will vary in their ease of proper placement in a 2D image, and image quality factors are likely to affect landmarks differently and in a complex way. High contrast, for example, may render a landmark more easily visible while, paradoxically, obscuring another. Pose angle is likely to be a further influential factor on the accuracy and precision of landmark placement in 2D.

References

Farkas, L. G., 1994. *Anthropometry of the head and face*, 2nd ed. New York: Raven Press.
R., 2008. *The R project for statistical computing*. http://www.r-project.org (accessed March 30, 2009).

Effect of 3D Rotation on Landmark Visibility

LORNA GOODWIN, DAMIAN SCHOFIELD, MARTIN PAUL EVISON, AND EDWARD LESTER

Contents

6.1 Introduction

To be useful in forensic facial comparison, anthropometric landmarks must be capable of distinguishing between subject's faces. These measurements must also offer sufficient repeatability in placement that observer error would not be significant when compared to variability due to face shape differences.

A set of 62 3D anthropometric landmarks were ranked according to their repeatability, observer error, and power to distinguish between subject's faces in Chapter 3. Landmark variation and observer error in 2D images of subjects' faces were described in Chapter 5. However, it is also necessary to address the issue of landmark visibility under different viewing conditions.

Images encountered in forensic facial comparison arise from a variety of sources, and the camera angle is often not chosen for the benefit of identification. This chapter is not an attempt to measure landmark visibility from these diverse sources; rather, it is a general investigation of landmark visibility in 3D, which provides potential guidelines for camera positioning for the purposes of identification.

6.2 Method of Analysis

Three subjects were scanned using the Cyberware® 3030PS Head and Neck Scanner, which unlike the Geometrix FaceVision® system, captures the full surface of the head, including the ears. The 3D head geometry for each scanned

Table 6.1 The 30 Landmarks Used in the Study

Landmark Number	Landmark	Label
1	Glabella	g
2	Sublabiale	sl
3	Pogonion	pg
4	Endocanthion left	en l
5	Endocanthion right	en r
6	Exocanthion left	ex l
7	Exocanthion right	ex r
8	Center point of pupil left	p l
9	Center point of pupil right	p r
10	Palpebrale inferius left	pi l
11	Palpebrale inferius right	pi r
12	Subnasion	se
13	Alar crest left	ac l
14	Pronasale	prn
15	Alar crest right	ac r
16	Highest point of columella prime left	c′ l
17	Highest point of columella prime right	c′ r
18	Labiale superius	ls
19	Labiale inferius	li
20	Stomion	sto
21	Cheilion left	ch l
22	Cheilion right	ch r
23	Superaurale left	sa l
24	Superaurale right	sa r
25	Subaurale left	sba l
26	Subaurale right	sba r
27	Postaurale left	pa l
28	Postaurale right	pa r
29	Otobasion inferius left	obi l
30	Otobasion inferius right	obi r

subject was imported into 3ds MAX® modeling software (Autodesk®, San Rafael, CA) using a 3ds file format.

A set of 30 landmarks was chosen for analysis (Table 6.1). These landmarks are the optimal set identified in Chapter 3, without substitution of the alares (al l and al r) for the alar crests (ac l and ac r). The landmarks were manually located and marked, using tools within the modeling software, on the 3D head geometry. Following the work of Aung et al. (1995), the nasal landmarks were located in columella view, with the head tilted back about 30°.

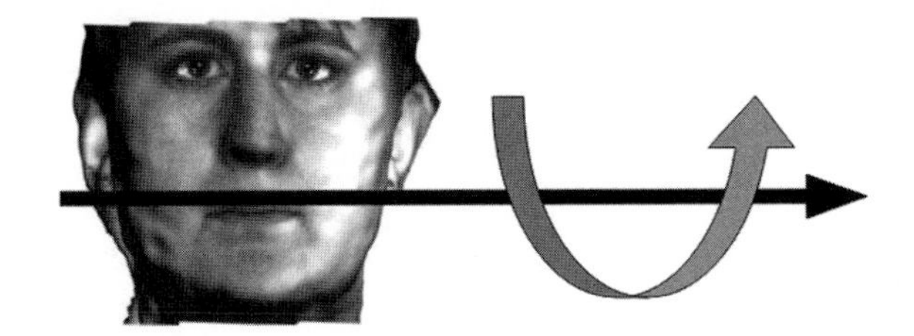

Figure 6.1 Pitch.

After the remaining landmarks had been placed, the head was orientated into consistent planes in each axis. The 3D head geometry was manually aligned into a consistent plane in 3ds MAX. Using the midline landmarks, the facial midline was orientated vertically in the coronal (x–y) plane from the front viewport window and in the transverse (x–z) plane from the lateral viewport window. The pronasale (prn) was not used, as this was observed to move off the midline in many individuals.

In the front viewport, the position of the endocanthions (en l and en r) and exocanthions (ex l and ex r) can be used for guidance as a line through these points (an interorbital line) will be at approximately 90° to the midline. These landmarks, and the superaurales (sa l and sa r) and subaurales (sba l and sba r), can also be used in the lateral view to provide similar guidance. The position of these landmarks cannot be expected to correspond perfectly with the midline as the head is not a symmetric geometric form.

Landmark visibility was assessed in relation to the three axes of head rotation: pitch, roll, and yaw (see Figures 6.1 to 6.3). Pitch (Figure 6.1) describes rotation around the x axis. Roll (Figure 6.2) describes rotation around the y axis. Yaw (Figure 6.3) describes rotation around the z axis.

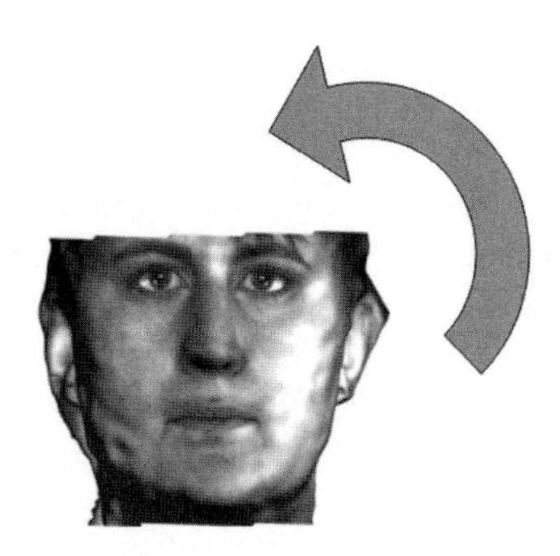

Figure 6.2 Roll.

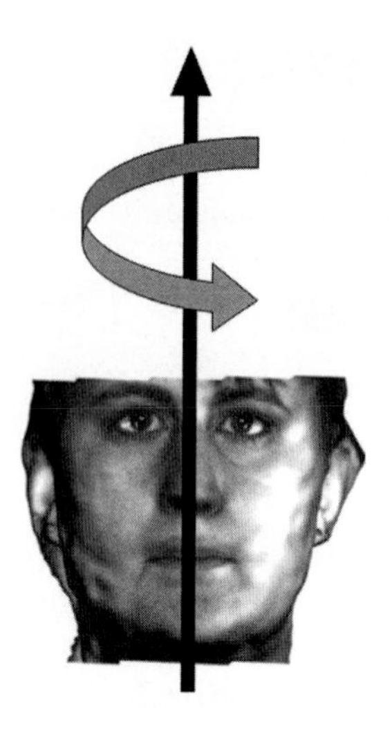

Figure 6.3 Yaw.

With the 3D geometry of each head consistently orientated into a starting position, it is now necessary to locate a consistent pivot point (this is often also described as the camera target or focal point) through which geometry might be rotated around the *x*, *y*, and *z* axes. There is no precise anatomical pivot point for a human head. An approximation that can be used as a point about which the living head rotates is a point at the intersection of the midsagittal section and a line from the otobasion inferius left (obi l) and otobasion inferius right (obi r) landmark points.

Although it is possible to rotate the 3D head geometry through 360° with each of the three axes (if increments of 1° were used), this would result in an impractical 360^3 or 46,656,000 possible orientations. For pragmatic reasons, therefore, 10° increments in orientation were chosen for measurement.

Consideration of the influence of pitch, roll, and yaw on landmark position and visibility indicates that pitch and yaw are likely to have significant influence, but that the influence of roll (Figure 6.2) is limited, in anterior view and in other combinations of pitch and roll. For this reason, roll was excluded from this analysis. Finally, few if any landmarks are visible from the rear views of the head, and it might be anticipated that practically, such views would not be used in forensic facial comparison. Therefore, ranges of pitch and roll of between −90° to +90° pitch and −90° to +90° yaw, from the start position, were chosen for use in the analysis.

Automation of reorientation of 3D head geometry between −90° to +90° pitch and −90° to +90° yaw, in 10° increments, was achieved using software

Table 6.2 3ds MAX Image Capturing Feature Settings and Explanations

Feature	Required Setting	Explanation
Material mapping	Self-illumination must be applied to each 3D landmark	Prevents shadows and light attenuation, hence preventing color alteration
	Diffuse, ambient, specular, and self-illumination RGB values to be identical	To enable "pure" RGB pixel color
Environment	Environmental effects, tints, and lighting should all be turned off	Prevents color alteration
Rendering	Images should be rendered at high resolution, preferably a minimum of 1000 × 750	Provides sharper edges and a greater number of pixels with which to view the 3D landmarks
	Antialiasing must be turned off	Prevents edge blurring
	Filter maps must be turned off	Prevents edge blurring
	Images should not be compressed	Prevents blurring and color alteration

developed within the 3ds Max modeling package, using the internal programming language, MAXScript®. In addition to automatic reorientation of the geometry, MAXScript was also used to render the reoriented images as frame views, and to recognize and assess the pixel color (RGB) values and identify the position of any pixel rendered in the frame.

In order to allow any pixel representing an anthropometric landmark to be distinguished, a uniform RGB value was used to render the head geometry and scene background, and 30 other distinct RGB values were used to render each of the 30 landmarks. In any orientation, therefore, the presence or absence (visibility) of any landmark can be automatically detected by the software.

Although the 3ds Max modeling package offers a complex virtual reality modeling and visualization environment, careful consideration of its parameters are essential to address potential sources of error when using such a 3D modeling and rendering approach. A list of modeling configuration parameters and rendering settings and the reason for these selections is shown in Table 6.2.

There are additional factors to be considered regarding compression of individual frames rendered from 3ds Max using AVI and MPEG codecs. Compression may distort pixel RGB values being used to assess landmark visibility. For this reason, for each of the three subjects scanned, 19 avi files were generated, one for each 10° increment of pitch, each containing 19 uncompressed frames, one for each 10° increment of yaw, from which the landmark visibility values could be derived (see Figure 6.4).

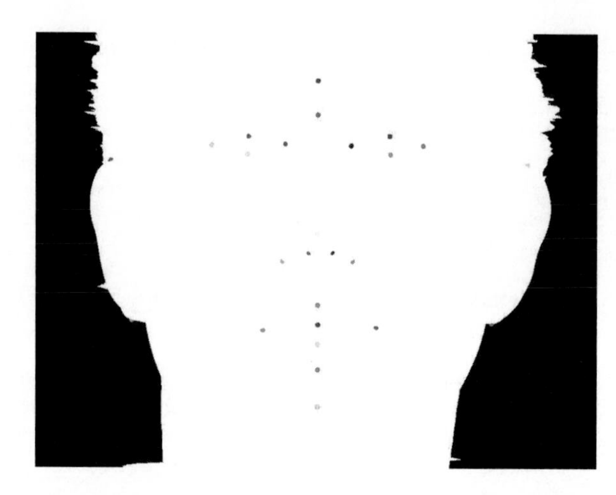

Figure 6.4 (See color insert following page 146.) Frame rendered in 3ds MAX showing visible landmarks in grayscale.

6.3 Results of Landmark Visibility Analysis

In order to allow convenient visualization of landmark visibility, the results for each landmark were collated to Microsoft® Office 2003 Excel. A Visual Basic program was written to format the data into a spreadsheet of 19 pitch and 19 yaw values, with landmark visibility shown as a color value representing true or false in each cell of the spreadsheet. A number of the 90 Microsoft Office 2003 Excel plots produced to show each landmark visibility spreadsheet for each subject are shown, for illustration, in Figures 6.5 to 6.12.

	Yaw Angle (left to right)																		
Pitch Angle	-90	-80	-70	-60	-50	-40	-30	-20	-10	0	10	20	30	40	50	60	70	80	90
-90																			
-80																			
-70																			
-60																			
-50																			
-40																			
-30																			
-20																			
-10																			
0																			
10																			
20																			
30																			
40																			
50																			
60																			
70																			
80																			
90																			

Figure 6.5 (See color insert following page 146.) Landmark visibility plot for the pronasale (prn) of subject 1. The landmark is visible at all angles of pitch and yaw analyzed.

	Yaw Angle (left to right)																		
Pitch Angle	−90	−80	−70	−60	−50	−40	−30	−20	−10	0	10	20	30	40	50	60	70	80	90
−90																			
−80																			
−70																			
−60																			
−50																			
−40																			
−30																			
−20																			
−10																			
0																			
10																			
20																			
30																			
40																			
50																			
60																			
70																			
80																			
90																			

Figure 6.6 (See color insert following page 146.) Landmark visibility plot for the stomion (sto) of subject 1.

Table 6.3 shows the ranking of landmarks by visibility calculated as the number of frames in which the landmark is visible (out of a possible 361). The ranking in power to distinguish between subject's faces (see Chapter 3) is also shown.

	Yaw Angle (left to right)																		
Pitch Angle	−90	−80	−70	−60	−50	−40	−30	−20	−10	0	10	20	30	40	50	60	70	80	90
−90																			
−80																			
−70																			
−60																			
−50																			
−40																			
−30																			
−20																			
−10																			
0																			
10																			
20																			
30																			
40																			
50																			
60																			
70																			
80																			
90																			

Figure 6.7 (See color insert following page 146.) Landmark visibility plot for the left endocanthion (en l) of subject 1. The endocanthion is visible at the smallest number angles of pitch and yaw analyzed.

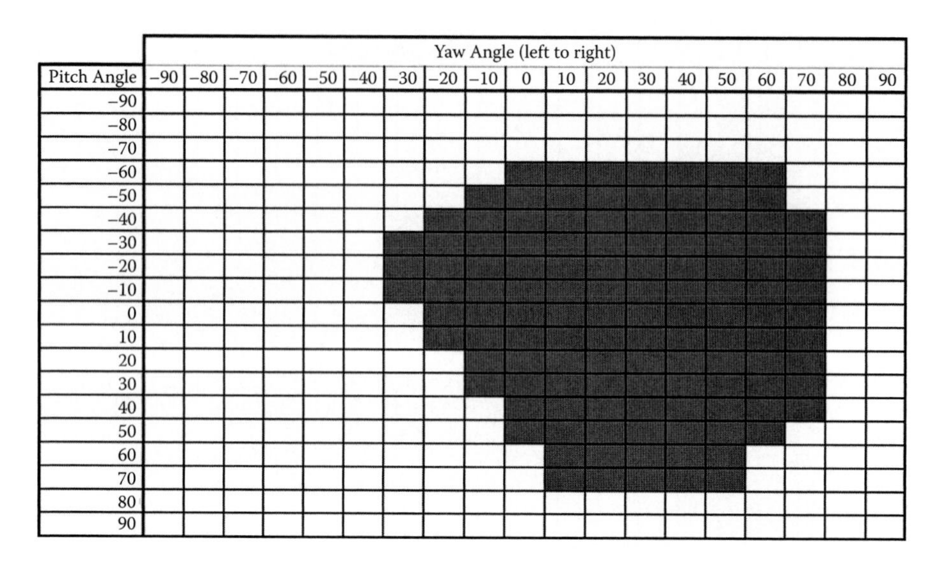

Figure 6.8 (See color insert following page 146.) Landmark visibility plot for the right endocanthion (en r) of subject 1. The endocanthion is visible at the smallest number angles of pitch and yaw analyzed.

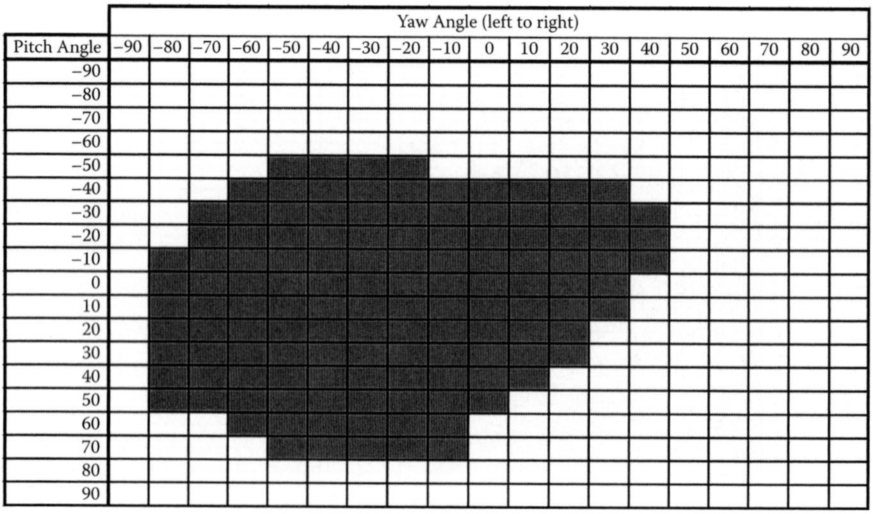

Figure 6.9 (See color insert following page 146.) Landmark visibility plot for the left endocanthion (en l) of subject 2. The endocanthion is visible at the smallest number angles of pitch and yaw analyzed.

	Yaw Angle (left to right)																		
Pitch Angle	−90	−80	−70	−60	−50	−40	−30	−20	−10	0	10	20	30	40	50	60	70	80	90
−90																			
−80																			
−70																			
−60																			
−50																			
−40																			
−30																			
−20																			
−10																			
0																			
10																			
20																			
30																			
40																			
50																			
60																			
70																			
80																			
90																			

Figure 6.10 (See color insert following page 146.) Landmark visibility plot for the right endocanthion (en r) of subject 2. The endocanthion is visible at the smallest number angles of pitch and yaw analyzed.

	Yaw Angle (left to right)																		
Pitch Angle	−90	−80	−70	−60	−50	−40	−30	−20	−10	0	10	20	30	40	50	60	70	80	90
−90																			
−80																			
−70																			
−60																			
−50																			
−40																			
−30																			
−20																			
−10																			
0																			
10																			
20																			
30																			
40																			
50																			
60																			
70																			
80																			
90																			

Figure 6.11 (See color insert following page 146.) Landmark visibility plot for the highest point of columella prime left (c′ l) of subject 1.

	Yaw Angle (left to right)																		
Pitch Angle	−90	−80	−70	−60	−50	−40	−30	−20	−10	0	10	20	30	40	50	60	70	80	90
−90																			
−80																			
−70																			
−60																			
−50																			
−40																			
−30																			
−20																			
−10																			
0																			
10																			
20																			
30																			
40																			
50																			
60																			
70																			
80																			
90																			

Figure 6.12 (See color insert following page 146.) Landmark visibility plot for the highest point of columella prime right (c′ r) of subject 1.

6.4 Summary

Landmark visibility follows a simple trend in which the prominence of facial features in any given combination of pitch and yaw is the most influential factor. Landmarks tend to remain visible until they become obscured by a more prominent feature than the one on which they are located.

The eight most visible landmarks are all central to the face, located on or near the facial midline. They tend not to be obscured by other facial features during pitch and yaw of the head. The pronasale (prn) was the only landmark visible over the entire range of head orientations for all three subjects (see Figure 6.5). The stomion (sto) tends to be visible at the maximum range of yaw, but becomes obscured, presumably by the upper and lower lip, when the head is pitched down or up, respectively (see Figure 6.6).

The endocanthions (en) have the least visibility for heads rotating between ±90° pitch and yaw (see Figures 6.7 to 6.10). They are easily hidden by surrounding facial features such as the nose and bridge of the nose, the brow ridges and the protruding eyeballs. In contrast, the highest point of columella prime (c′) landmarks are obscured by the nasal tip and alares in pitch and the cheeks in yaw (see Figures 6.11 and 6.12).

The pronasale (prn) and other midline landmarks do not line up perfectly as the face is not always symmetrical. Furthermore, there is often variation between individuals, as illustrated by the comparison of the endocanthions (en) of subjects 1 and 2 (see Figures 6.7 to 6.10).

Table 6.3 Landmarks in Order of Visibility Level with the Ranking in Power to Distinguish between Faces (see Chapter 3) Also Shown

	Number of Frames Visible				Ranking in Power to Distinguish between Subjects' Faces
Landmark	Subject 1	Subject 2	Subject 3	Total	
prn	361	361	361	1083	1
pg	356	356	353	1065	3
li	349	349	346	1044	22
ls	347	341	349	1037	33
g	316	355	355	1026	39
se	330	326	335	991	42
sl	280	320	326	926	11
sto	275	287	244	806	26
c′ r	234	256	260	750	18
c′ l	245	247	240	732	19
pi r	226	231	239	696	13
pi l	227	233	226	686	12
p r	217	220	232	669	10
p l	216	211	213	640	8
ex r	202	199	222	623	25
ex l	197	199	209	605	15
al r	192	216	195	603	55
al l	180	210	206	596	49
ch l	136	228	214	578	17
ch r	138	209	219	566	20
sba r	188	177	196	561	7
obi r	184	180	192	556	16
sba l	173	185	184	542	4
obi l	176	178	187	541	21
sa r	160	165	169	494	9
pa r	185	156	136	477	5
pa l	175	169	108	452	2
sa l	143	147	159	449	6
en l	145	127	119	391	28
en r	121	137	131	389	27

After the endocanthions (en), the next least visible landmarks are located on the ears, closely followed by the cheilions (ch) on the corners of the mouth. As the ear landmarks are farthest from the front of the face, they will be visible within limited ranges of yaw, and become quickly obscured by hair, the ear itself, and the other facial features. The cheilions (ch) are easily hidden by the mouth, as they protrude from the face very little.

The remaining landmarks fall somewhere in between the pronasale (prn) and endocanthions (en), depending on their position and potential to be obscured by other facial features during pitch and yaw of the head, as reflected in their position in Table 6.3, according to the following factors:

- The level of protrusion of the landmark: the distance between the landmark and the pivot point.
- The proximity, position, shape, and relative size of other obscuring features

Two wider factors will inevitably have a major influence:

- Variation in these factors from individual to individual
- Camera position

Individual variation will influence landmark protrusion and the ability of other facial features to obscure their visibility. In this investigation, the pronasale (prn) was chosen as the camera focal point. The pattern of visibility would have been different if another focal point had been chosen.

There is no obvious relationship between landmark visibility and distinguishing power (see Table 6.3). Further research, however, may permit the position of a camera or cameras to be identified, which will yield optimum landmark visibility—or the best combination of visibility and distinguishing power. The approach used in this investigation would readily lend itself to:

- A complete analysis of 360° of pitch, roll, and yaw
- An analysis with each landmark chosen as the focal point
- Analysis of a larger sample

Finally, there is the potential to "reverse engineer" the approach used in this investigation to a more general question relevant to forensic facial comparison: should these landmarks be visible in this pose for this subject? If the answer is "no," for example, an exclusion could possibly be made.

References

Aung, S. C., R. C. K. Ngim, and S. T. Lee, 1995. Evaluation of the laser scanner as a surface measuring tool and its accuracy compared with direct facial anthropometric measurements. *Br. J. Plast. Surg.* 48(8): 527–621.

Influence of Lens Distortion and Perspective Error

7

DAMIAN SCHOFIELD, MARTIN PAUL EVISON, AND LORNA GOODWIN

Contents

7.1 Introduction

In an anthropometric landmark based approach to forensic facial comparison, a 2D image of an offender typically will be compared with a 2D or, potentially, 3D image of a suspect.

Stereophotography (see Chapter 3) offers a precise means of locating anthropometric landmarks in 3D. The origin of 2D images is likely to be diverse and could arise from a variety of analog or digital photographic or video sources. The sources of potential error in accurate landmark location are similarly diverse, although one factor—that of lens distortion—has the potential to affect all systems.

Aberrations in optical systems were the subject of early research in the field (for an introduction, see Ray 2002). Distortion, in particular, is defined as the displacement of an image point from where it should be if the object plane were mapped at a constant magnification on to the image plane. Displacement is typically curvilinear. If the point is displaced inward (toward the optical axis) this is known as *pin cushion distortion*, if the point is displaced outward (away from the optical axis) this is known as *barrel distortion* (see Figure 7.1).

There are two forms of nonoptical distortion that will affect the position of landmarks in a 2D image. Perspective distortion is a form of nonoptical distortion, where nearby subjects are rendered larger than faraway subjects of the same size, leading to a sense of depth or convergence of lines that are

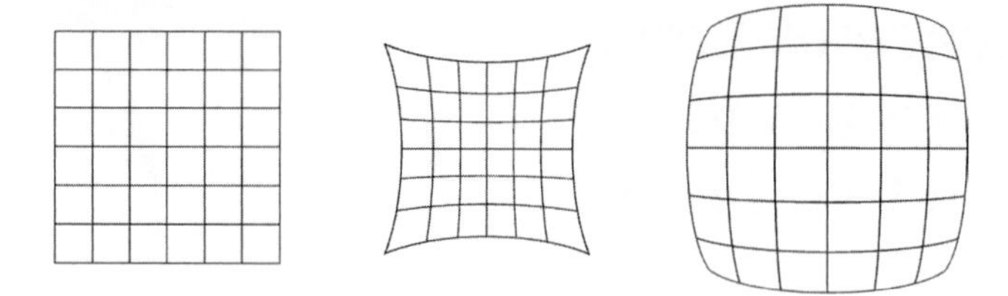

Figure 7.1 Distortion of a rectangular grid. Left: undistorted grid; center: pincushion distortion; right: barrel distortion.

parallel. Perspective is not affected by the lens focal length or camera distance: it depends on the perspective projection as determined by the viewpoint of the camera in relation to the subject.

Geometric distortion results when a 3D object is projected onto a flat plane. A well known example occurs when a sphere at the center of an image is rendered as a round disk on the film, while a sphere at the periphery of an image is elliptically elongated.

This chapter describes an investigation of the potential influence of lens distortion on landmark positioning, with a view to developing guidelines for 2D image capture and measurement, and—in particular—to determining the distance that a camera has to be from the subject before the perspective view matches the orthogonal projection.

7.2 Method of Analysis

An investigation of the effects of lens distortion on manual landmarks placed on 2D images in a sample of live subjects captured with a variety of different lenses, while controlling for perspective distortion, would present a prohibitively laborious and time-consuming challenge. For this reason, an *in silico* approach using visualization modeling software was chosen, permitting automation of much of the analytical process.

Twelve 3D images were provisionally assessed, which represented extremes of y and z (width and depth, respectively) dimensions in the 3D image database collected using the Geometrix FaceVision® FV802 Series Biometric Camera (ALIVE Tech, Cumming, GA). From this set of 12, five 3D images were selected which showed the fewest scanner anomalies (see Chapter 2).

The five 3D images were landmarked at 30 sites (see Table 7.1) in the 3ds Max® modeling software (Autodesk®, San Rafael, CA).

After the remaining landmarks had been placed, the head was orientated into consistent planes in each axis. The 3D head geometry was manually

Table 7.1 The 30 Landmarks Used in the Study

Landmark Number	Landmark	Label
1	Glabella	g
2	Sublabiale	sl
3	Pogonion	pg
4	Endocanthion left	en l
5	Endocanthion right	en r
6	Exocanthion left	ex l
7	Exocanthion right	ex r
8	Center point of pupil left	p l
9	Center point of pupil right	p r
10	Palpebrale inferius left	pi l
11	Palpebrale inferius right	pi r
12	Subnasion	se
13	Alar crest left	ac l
14	Pronasale	prn
15	Alar crest right	ac r
16	Highest point of columella prime left	c′ l
17	Highest point of columella prime right	c′ r
18	Labiale superius	ls
19	Labiale inferius	li
20	Stomion	sto
21	Cheilion left	ch l
22	Cheilion right	ch r
23	Superaurale left	sa l
24	Superaurale right	sa r
25	Subaurale left	sba l
26	Subaurale right	sba r
27	Postaurale left	pa l
28	Postaurale right	pa r
29	Otobasion inferius left	obi l
30	Otobasion inferius right	obi r

aligned into a consistent plane in 3ds Max. Using the midline landmarks, the facial midline was orientated vertically in the coronal (x–y) plane from the front viewport window and in the transverse (x–z) plane from the lateral viewport window. The pronasale (prn) was not used, as this was observed to move off the midline in many individuals.

In the front viewport, the position of the endocanthions (en l and en r) and exocanthions (ex l and ex r) can be used for guidance as a line through these points (an interorbital line) will be at approximately 90° to the midline. These landmarks, and the superaurales (sa l and sa r) and subaurales (sba l and sba r), can also be used in the lateral view to provide similar guidance. Similarly, the landmarks of the circumoral or mouth region—the labiale superius (ls), labiale inferius (li), stomion (sto), and cheilion (ch)—were located using lateral, anterior, and user-selected views, where left and right bilateral landmarks were again checked for plane alignment. The position of these landmarks cannot be expected to correspond perfectly with the midline as the head is not a symmetric geometric form.

The landmarks are the optimal set identified in Chapter 3, without substitution of the alares (al l and al r) for the alar crests (ac l and ac r). The landmarks were located manually on the 3D head geometry in 3ds MAX. Following Aung et al. (1995), the nasal landmarks were located in columella view—with the head tilted back about 30°—after the remaining landmarks had been placed and the head orientated into a consistent plane.

The landmarks were assigned individual, distinct RGB colors. Original landmark geometry was measured by algorithmic calculation (using software written in the MAX Script® programming language) of pairwise distances between landmarks, yielding 435 pairwise distances between the 30 sites.

The influence of lens distortion was assessed *in silico* via rendering of 2D anterior views of the five subjects' 3D image datasets at different lens parameter settings in the 3ds Max modeling software. The pronasale (prn) approximates to the most central point of the face in the anterior view and was chosen as the focal point for the virtual camera.

The following parameters were used:

- Subject to camera distance: 0.5, 1.0, 1.5, or 2.0 m
- Lens: 15 mm, 35 mm or 50 mm (CCTV camera lenses usually range approximately from less than 10 mm up to 50 mm; 15 mm is the minimum offered in the modeling software)
- View: orthographic or perspective

Each of the 24 possible camera combinations for the five subjects was rendered in 2D in uncompressed bitmap format. A 4000 × 3000 pixel resolution was used, because in tests not all visible landmarks were rendered from the 15 mm lens at 2 m distance when lower resolutions were used. Similarly, it was established during testing that the OpenGL® rendering drivers yielded the best quality alignment of rendered textures to the original wireframe.

The pixel coordinates at the center of each landmark in the rendered image were manually measured and then recorded in Microsoft® Office 2003 Excel spreadsheets. If necessary, reference was made to the 3D image to ensure correct texture alignment and the proper location of the landmark.

The 435 pairwise distances between landmarks were calculated automatically using Microsoft Office 2003 Excel spreadsheet macros and programs written using Visual Basic. For each view, a pixel scaling factor in pixels per mm was calculated trigonometrically from the original 3D model in order to convert pixel distances to mm for both perspective and orthogonal views.

In order to assess perspective error, differences between pairwise distances measured from perspective and orthogonal projections were calculated for all 24 camera combinations. The minimum and maximum values, ranges, and standard deviation were also calculated.

Minimum and maximum values were also calculated for each pixel scaling factor, and for differences in pairwise pixel distances between landmarks measured in perspective and orthogonal views. These values were used to check for possible pixel measurement errors.

Perspective errors were calculated as the ratio of perspective distance to orthogonal distance between each pair of landmarks in pixels, converted into mm using the pixel scaling factor for any particular pairwise measurement.

7.3 Results

Figure 7.2 shows a plot of perspective error, calculated as the difference between pairwise distances in orthographic and perspective views, versus pairwise distance in 3D. It can be seen clearly from the figure that perspective error affects the landmarks of the ear to a greater extent than other landmarks.

Figures 7.3 to 7.7 show plots of perspective error in mm for each pairwise distance for each of the five subjects, in ascending order of 2D Euclidean distance between landmarks for 435 landmark pairs.

Figures 7.3 to 7.7 show clear overall patterns of results, irrespective of the subject, these are:

- Perspective errors decrease roughly exponentially with camera distance.
- There is a close correlation between perspective error and 2D pairwise distance.
- The relationship between perspective error and 2D pairwise distance begins as a roughly linear relationship and shows a sudden rise, which appears proportionally greater for smaller camera distances, and continues as a line or slope with a sudden rise at the final and longest distances, which again appears proportionally greater for smaller camera distances.

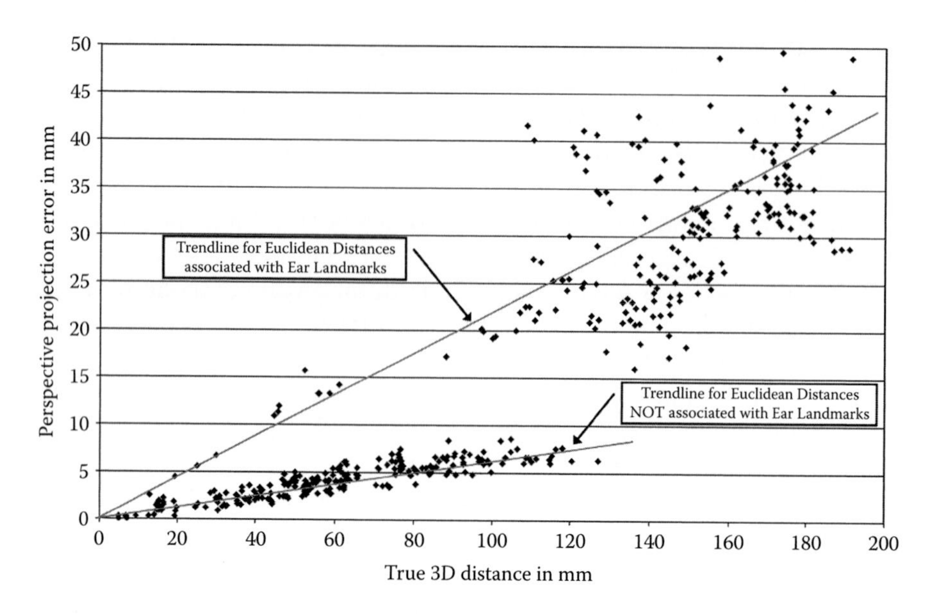

Figure 7.2 Chart showing the relationship between perspective error and "true" pairwise Euclidean distance between landmarks in 3D in mm for all 435 pairwise distances (lens: 15 mm; camera distance: 0.5 m). Note: separate trend lines are shown for ear and nonear landmarks.

Close inspection of Figures 7.3 to 7.7 indicates that there are some differences in the general influences of perspective error between individual subjects, however. This order of ascending error closely matches the order of ascending facial depth, as described in greater detail below, which suggests a relationship between the two factors.

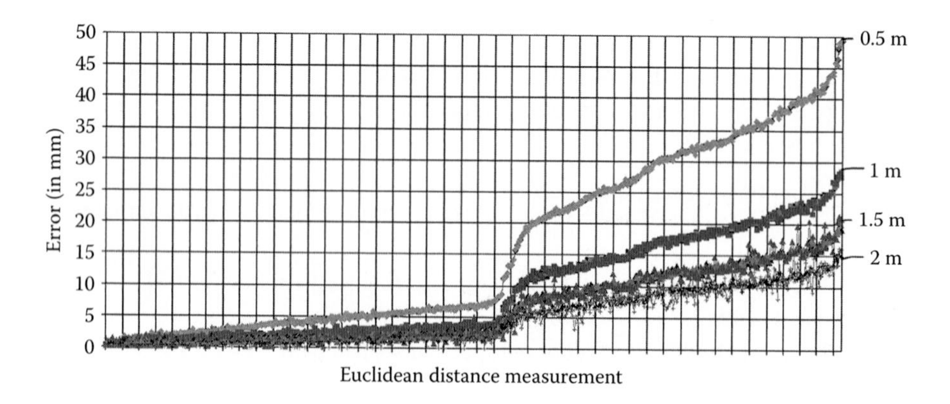

Figure 7.3 Plots for all four camera distances of average value and range of perspective error in mm for each of 435 2D pairwise measurements between landmarks, shown in ascending order of Euclidean distance (lens: 50 mm; subject one).

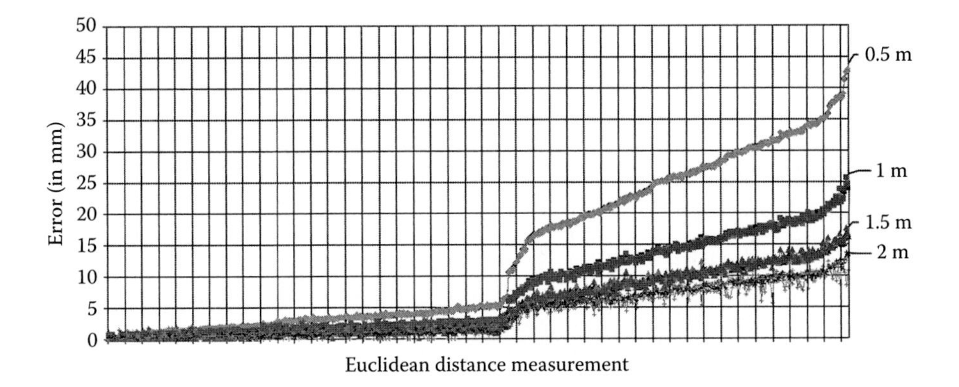

Figure 7.4 Plots for all four camera distances of average value and range of perspective error in mm for each of 435 2D pairwise measurements between landmarks, shown in ascending order of Euclidean distance (lens: 50 mm; subject two).

Figure 7.8 shows a plot of the percent ratio of perspective distance versus orthogonal distance for each landmark where no adjustment is made against the original 3D pairwise distance, for subject four only.

Figure 7.8 illustrates that when perspective error is not considered relative to the 3D pairwise distance, shorter measurements, such as around the mouth, eyes, or nose, show significantly higher perspective errors and perspective error ranges than others. This is because a small measurement error, for example, of ±1 pixel on each landmark coordinate, confers an apparent perspective error ratio of greater magnitude for orthogonal distances

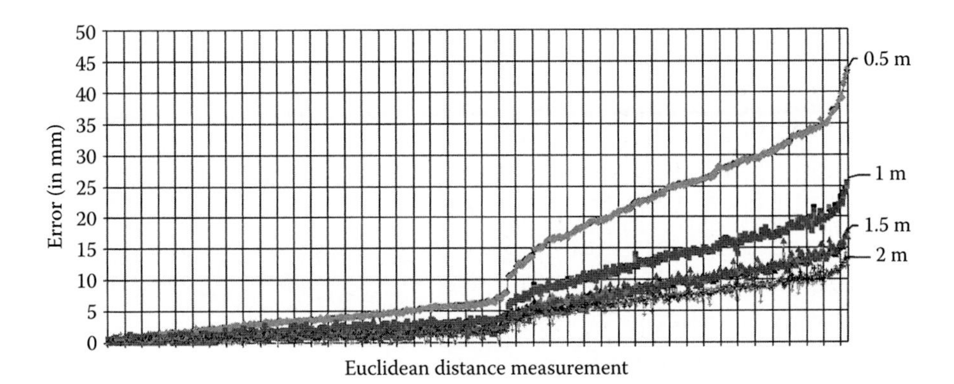

Figure 7.5 Plots for all four camera distances of average value and range of perspective error in mm for each of 435 2D pairwise measurements between landmarks, shown in ascending order of Euclidean distance (lens: 50 mm; subject three).

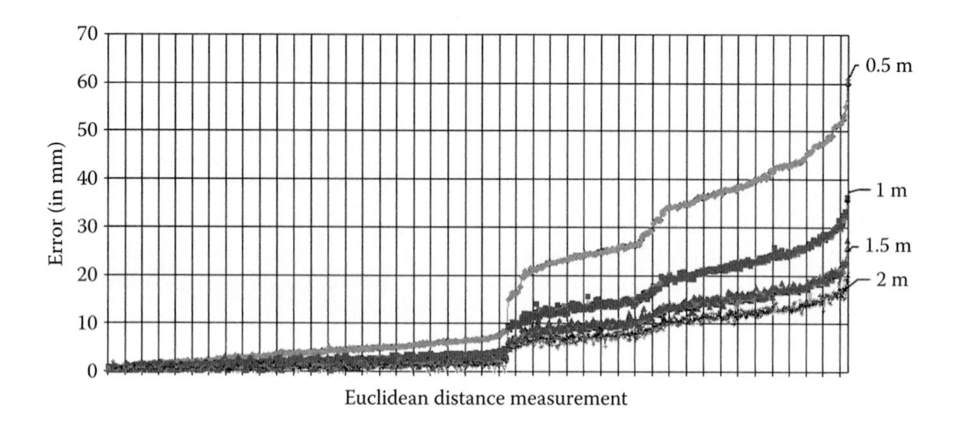

Figure 7.6 Plots for all four camera distances of average value and range of perspective error in mm for each of 435 2D pairwise measurements between landmarks, shown in ascending order of Euclidean distance (lens: 50 mm; subject four).

constituting a smaller number of pixels. Shorter distances are more subject to measurement error, but less subject to perspective error relative to longer ones.

Figure 7.9 shows an alternative plot form for 145 of the 435 pairwise comparisons, illustrating the measurement error effect on adjacent landmarks on the eyes, nose, mouth, and ears for subject three.

In addition to lens setting and camera distance, fundamental dimensionality, width, height, and depth, are anticipated to influence perspective error. Figures 7.10 to 7.14 show plots of perspective error ranking (see Figures 7.3

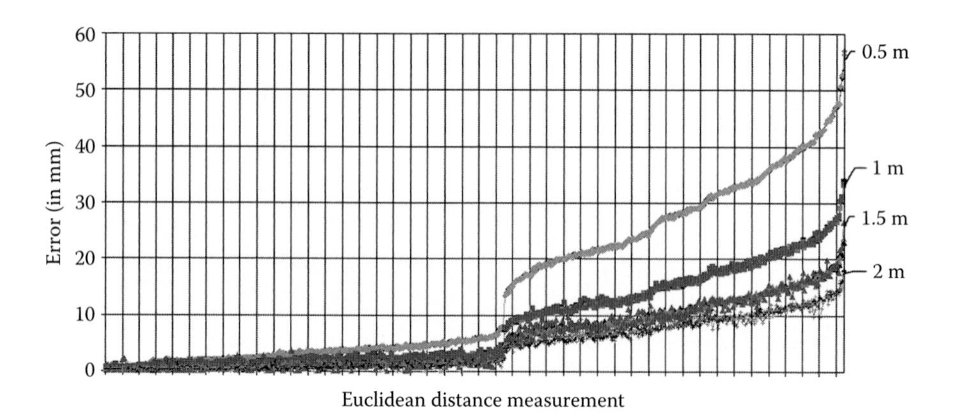

Figure 7.7 Plots for all four camera distances of average value and range of perspective error in mm for each of 435 2D pairwise measurements between landmarks, shown in ascending order of Euclidean distance (lens: 50 mm; subject five).

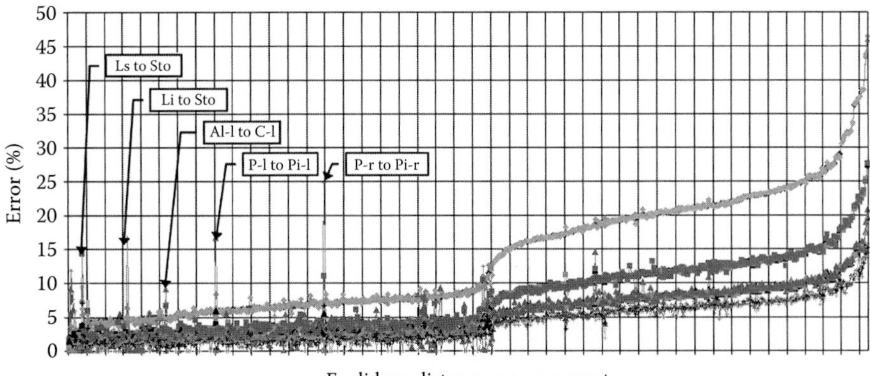

Figure 7.8 Plots for all four camera distances of average value and range of per cent perspective error for each of 435 2D pairwise measurements between landmarks, shown in ascending order of Euclidean distance (lens: 50 mm; subject four). Pairwise distances that are sources of higher error are labeled.

to 7.8) versus z-axis difference (or depth) for each pairwise measurement for subjects one to five.

Figures 7.10 to 7.14 illustrate the influence of dimensionality in the x and z dimensions, as well as the y dimension shown in the figures. It is possible to distinguish a general relationship between the gross dimensionality

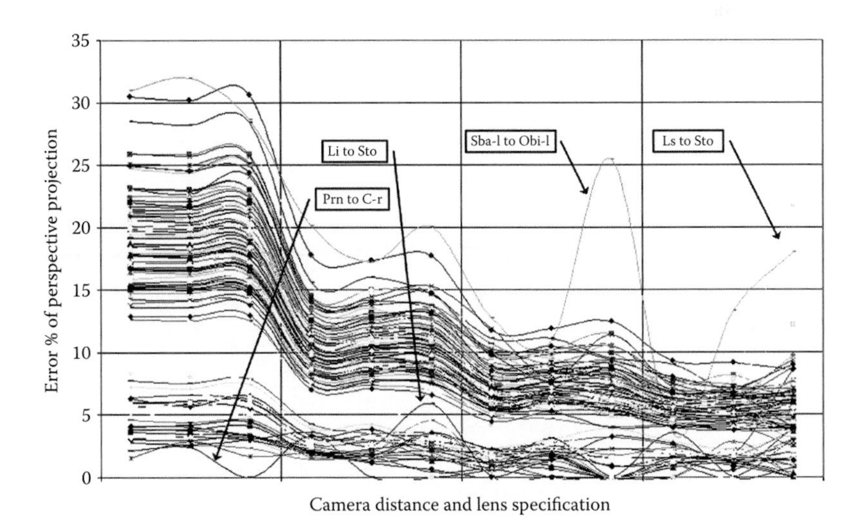

Figure 7.9 Plots for all four camera distances of average percent perspective for 145 of 435 2D pairwise measurements between landmarks, shown by lens setting within camera distance; the large steps affecting all plots are due to changes in camera distance. Pairwise distances that are sources of higher error are labeled.

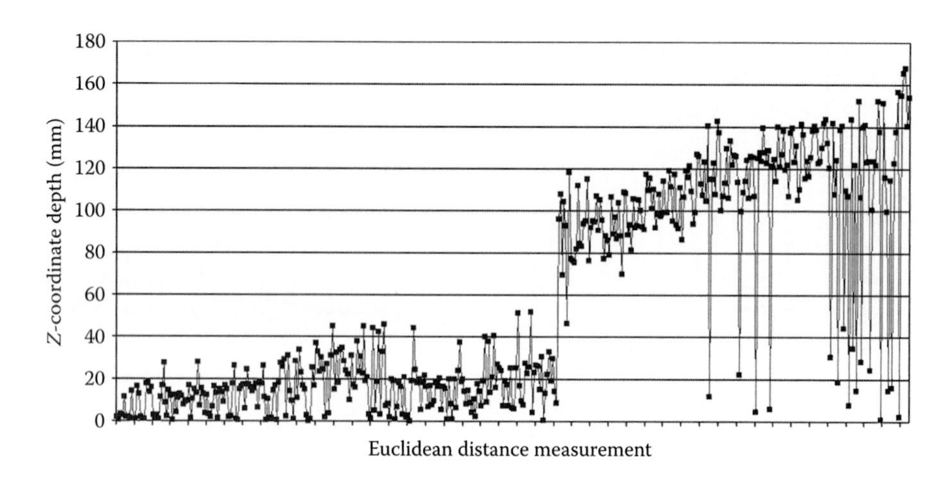

Figure 7.10 Plot of ascending rank order of perspective error for each of 435 2D pairwise measurements between landmarks against *z*-coordinate distance or depth for subject one.

(width, height, or depth) and the perspective error ranking, which could be anticipated given the finding of a correlation between perspective error and orthogonal distance. There are exceptions, however. The greatest *y*-coordinate distance (or height) peaks near the center of the perspective error ranking, with the glabella (g) to pogonion (pg) distance. This measurement is

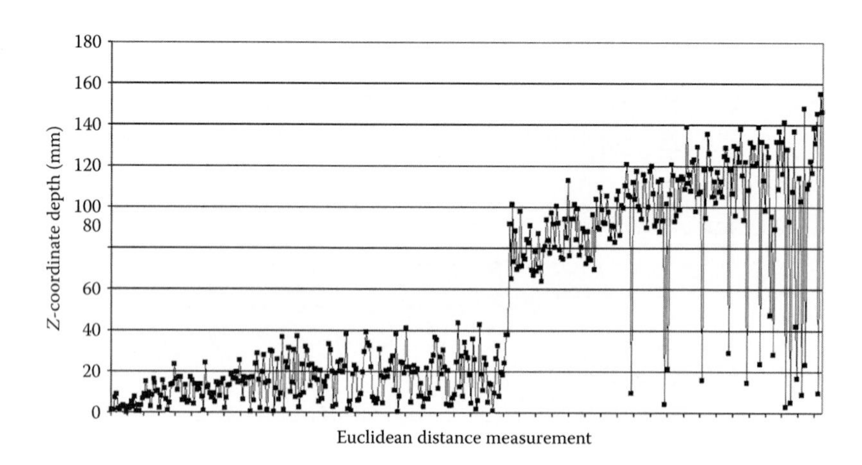

Figure 7.11 Plot of ascending rank order of perspective error for each of 435 2D pairwise measurements between landmarks against *z*-coordinate distance or depth for subject two.

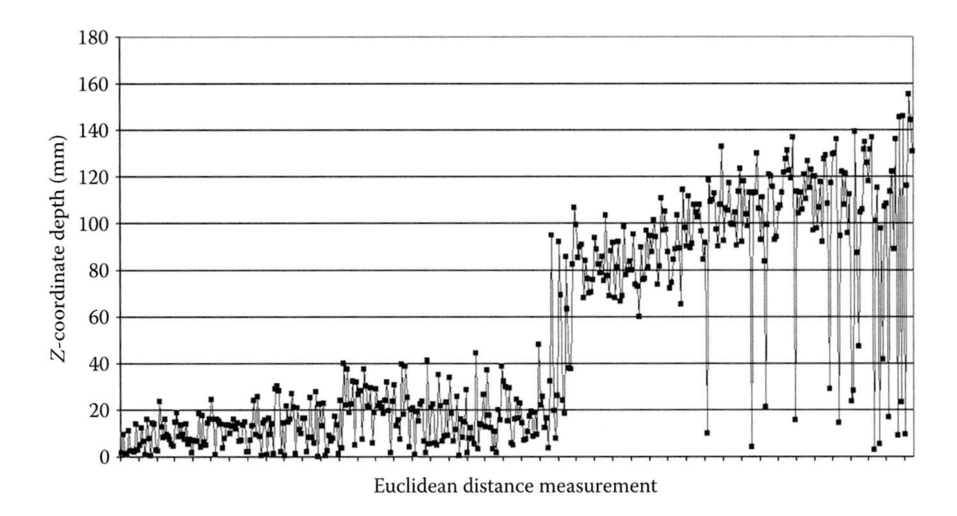

Figure 7.12 Plot of ascending rank order of perspective error for each of 435 2D pairwise measurements between landmarks against *z*-coordinate distance or depth for subject three.

noticeably close to the midline and camera focal point at the pronasale (prn) or nasal tip.

The *z*-coordinate distance (or depth) illustrates further patterns, which appear to split the Euclidean distances into two large groups (see Figures 7.10 to 7.14).

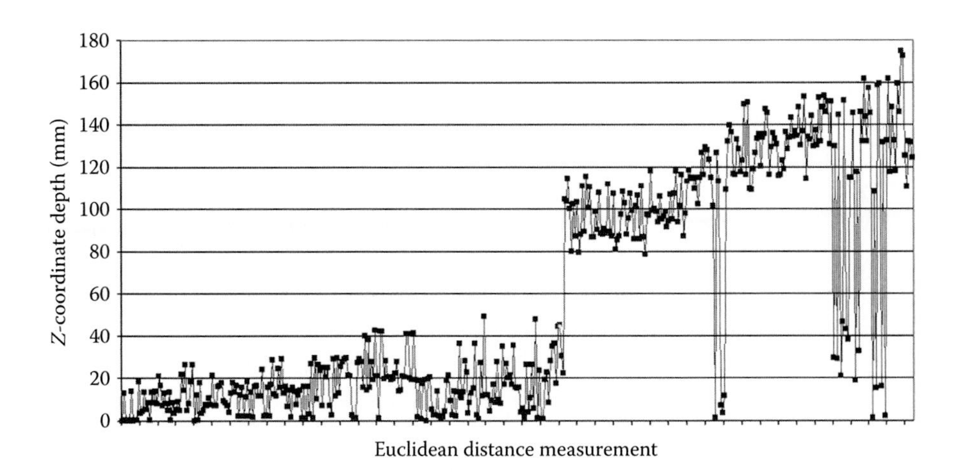

Figure 7.13 Plot of ascending rank order of perspective error for each of 435 2D pairwise measurements between landmarks against *z*-coordinate distance or depth for subject four.

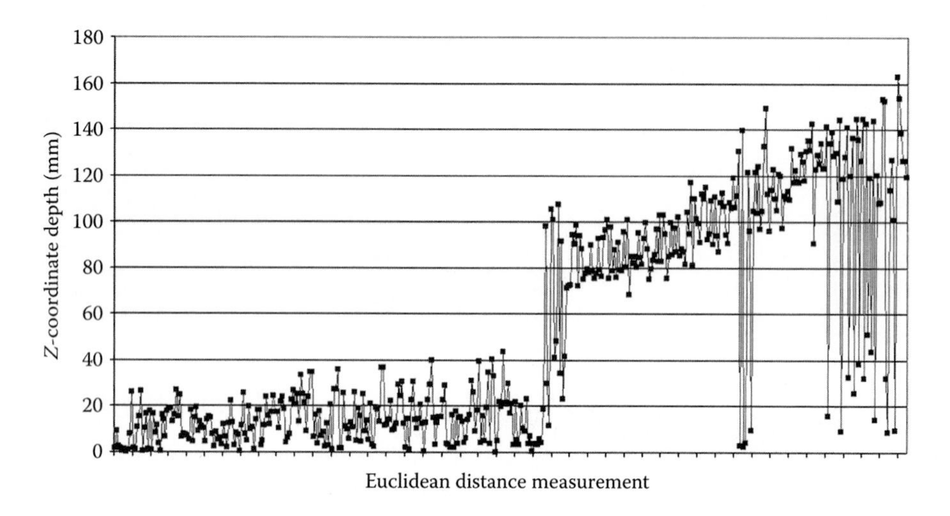

Figure 7.14 Plot of ascending rank order of perspective error for each of 435 2D pairwise measurements between landmarks against *z*-coordinate distance or depth for subject five.

In both groups, the pairwise distances with the highest *z*-dimensionality (i.e., those which span the greatest depth) rank highest in perspective error. The second group (incorporating the landmarks of the ear region) has markedly higher perspective error, but with some clear exceptions. These exceptions are pairwise distances between opposite ear region landmarks. Although these distances have little depth, they will be subject to perspective error correlated with dimensionality in other axes. The same pattern is evident for all five subjects, although there are general trends relating to the individual size and shape of each head. In Figures 7.10 to 7.14 it is variation in the *z* dimension or depth that is shown.

As a consequence of these observations, the influence of the distance between the landmarks and the camera focal point on the pronasale or nasal tip was investigated directly.

Figures 7.15 to 7.19 show plots of 3D pairwise distances from each landmark to the pronasale (prn) versus cumulative perspective error for pairwise distance for all five subjects. Combinations of the x, y, and z dimensionality (width, height, and depth) relating to each pairwise distance from the pronasale (prn) is also plotted.

Figures 7.15 to 7.19 show a very subtle increase in perspective error from those landmarks on the facial midline to those in the bilateral regions of the eyes and lips. There is a marked increase in perspective error when the

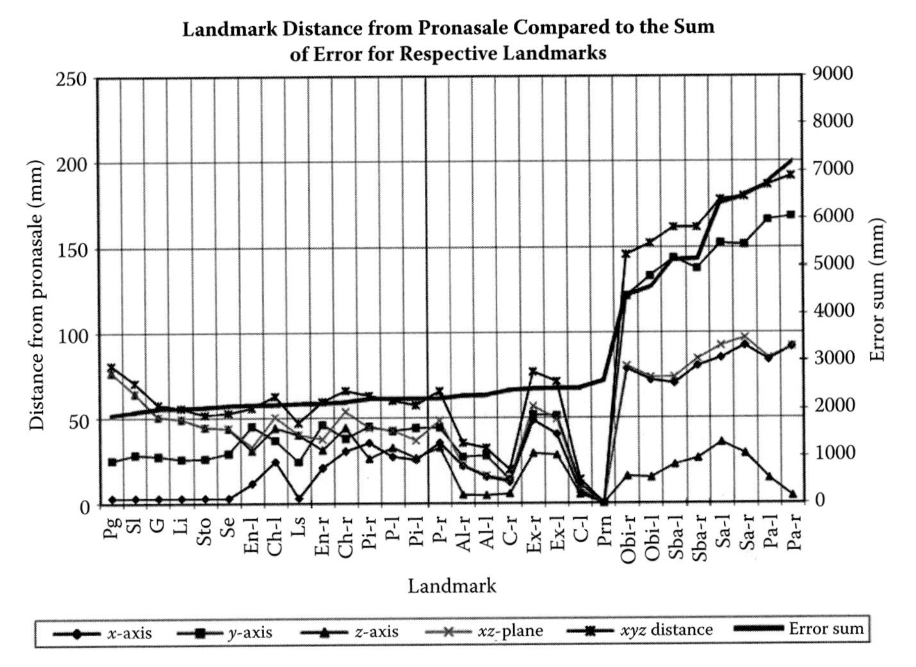

Figure 7.15 Chart showing plots of cumulative perspective error in pairwise distances from the pronasale (prn) to the other 29 landmarks, in *x*, *y*, *z*, *xz*, and *xyz* dimensions for subject one.

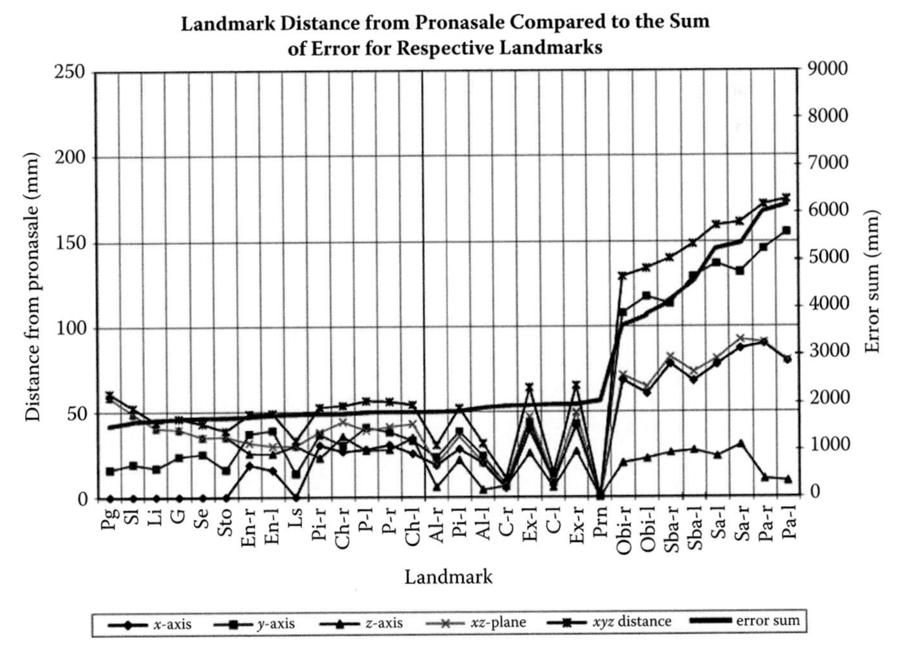

Figure 7.16 Chart showing plots of cumulative perspective error in pairwise distances from the pronasale (prn) to the other 29 landmarks, in *x*, *y*, *z*, *xz*, and *xyz* dimensions for subject two.

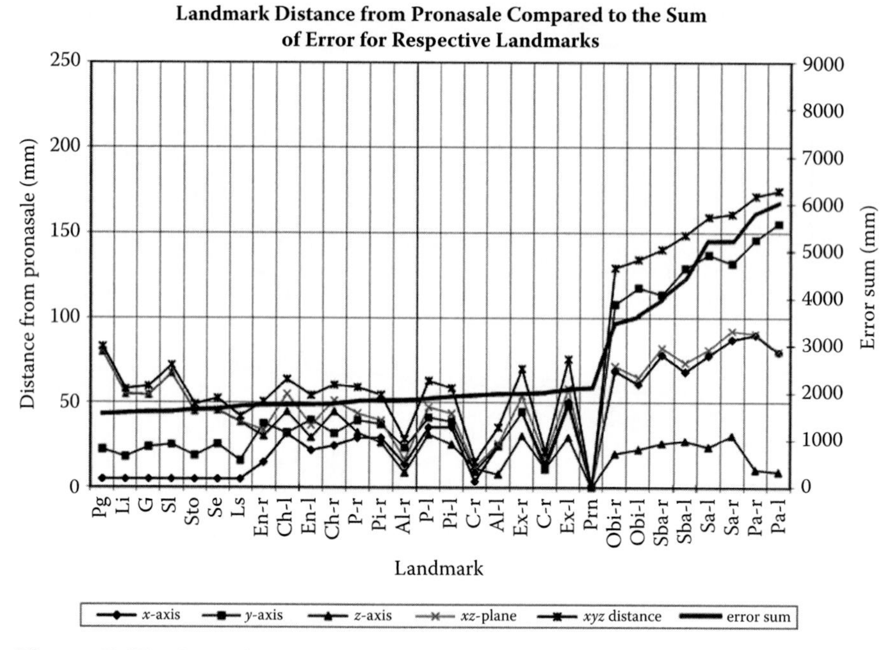

Figure 7.17 Chart showing plots of cumulative perspective error in pairwise distances from the pronasale (prn) to the other 29 landmarks, in *x*, *y*, *z*, *xz*, and *xyz* dimensions for subject three.

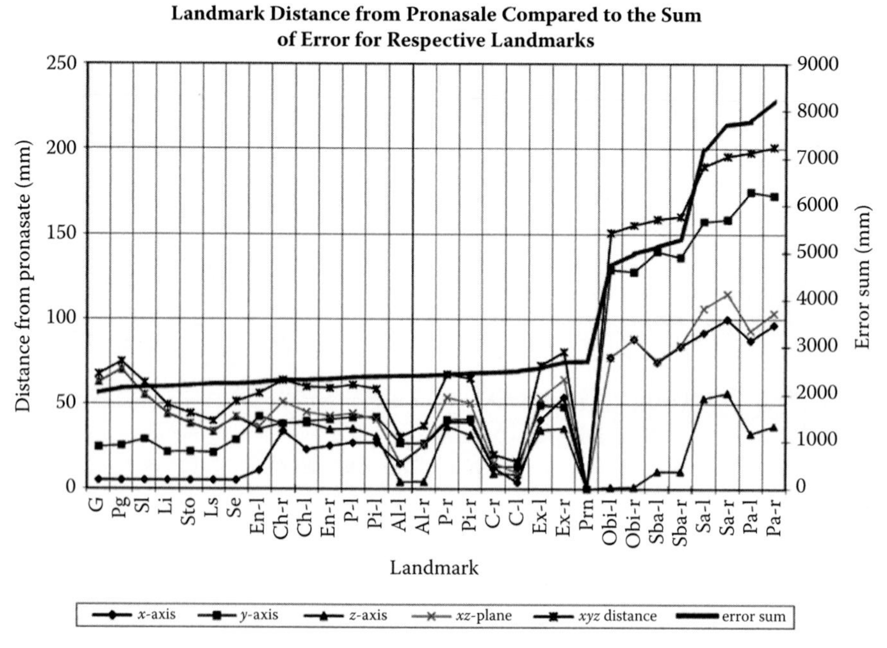

Figure 7.18 Chart showing plots of cumulative perspective error in pairwise distances from the pronasale (prn) to the other 29 landmarks, in *x*, *y*, *z*, *xz*, and *xyz* dimensions for subject four.

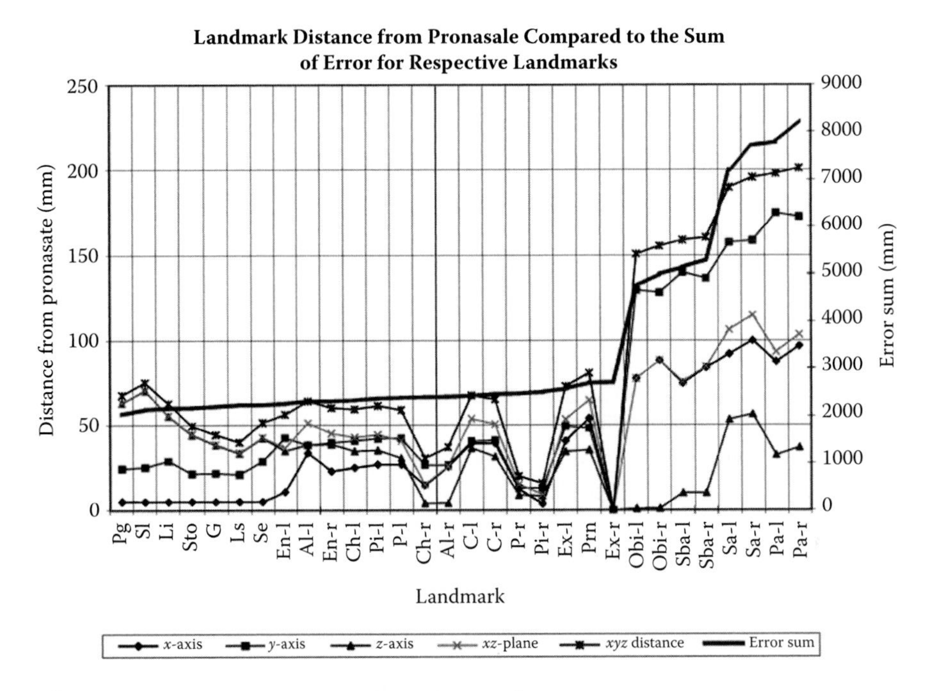

Figure 7.19 Chart showing plots of cumulative perspective error in pairwise distances from the pronasale (prn) to the other 29 landmarks, in *x*, *y*, *z*, *xz*, and *xyz* dimensions for subject five.

landmarks of the ear region are encountered. Here, an underlying relationship between 3D (*x*, *y*, *z*) distance and perspective error can be correlated with *z* dimensionality, or depth, and then *x* dimensionality, or width. The *y* dimension (height) does not appear to have noticeable influence.

It is tempting to anticipate that for the landmarks close to the pronasale (prn), lack of obvious correlation with fluctuations in dimensionality is due to very low perspective error being associated with landmarks subject to random measurement error (see above).

Table 7.2 illustrates the cumulative perspective error in mm calculated for each landmark for each subject. Note that this is derived form the pairwise distance errors associated with the landmark in combination with the other 29 pairings.

The ranking in Table 7.2 reflects the correlation between perspective error and pairwise distance. Although the pronasale (prn) is located at the camera focal point, which has no perspective error, it is also consistently distant from all its peripheral landmarks, leading to a cumulative perspective error correlating with those distances.

Table 7.2 Cumulative Perspective Error in mm for Each Landmark for Each Subject

	Cumulative Perspective Error in mm for Each Subject					
Landmark	1	2	3	4	5	Total
Pg	1859	1506	1557	2130	1553	8605
Sl	1931	1596	1603	2156	1759	9047
G	2001	1653	1601	2032	1838	9125
Li	2006	1630	1580	2166	1815	9196
Sto	2027	1678	1653	2193	1838	9387
Se	2061	1674	1660	2226	1907	9528
Ls	2103	1747	1725	2226	1860	9661
En-l	2076	1733	1758	2256	1924	9747
En-r	2110	1694	1746	2328	1986	9864
Ch-l	2082	1796	1750	2310	1997	9935
Ch-r	2137	1760	1767	2308	2021	9993
P-l	2207	1794	1880	2366	2020	10268
Pi-l	2209	1811	1926	2381	2017	10346
Al-r	2266	1798	1849	2401	2048	10361
P-r	2209	1795	1829	2423	2114	10370
Pi-r	2207	1756	1845	2455	2133	10396
Al-l	2277	1885	1980	2395	1972	10509
C-r	2383	1926	1952	2466	2109	10836
C-l	2421	1953	2009	2497	2082	10961
Ex-l	2421	1934	2081	2567	2184	11187
Ex-r	2411	1954	1984	2689	2350	11388
Prn	2586	2040	2112	2699	2238	11676
Obi-r	4361	3599	3471	4990	4290	20711
Obi-l	4563	3832	3632	4735	4086	20848
Sba-r	5150	4131	3997	5292	4683	23252
Sba-l	5118	4567	4439	5131	4434	23690
Sa-l	6314	5230	5214	7195	5587	29540
Sa-r	6476	5358	5219	7710	6545	31309
Pa-l	6733	6175	6043	7780	6724	33455
Pa-r	7201	6007	5788	8230	7573	34798

7.4 Summary

The influence of both lens distortion and perspective error on pairwise distances between landmarks have been investigated and, as would be anticipated, follow the classic model, the influence of which varies in reflection to the unique geometry of the human facial form. The subjects chosen in the investigation tended toward the extremes of overall dimensionality in the large sample database.

Observations are a combination of the effects of lens distortion, perspective error, and measurement error:

- There is an increase in perspective error correlated with distance from the camera focal point.
- There is an inverse exponential correlation between lens distortion and camera distance.
- Pairwise distance measurement error is inversely correlated with pairwise distance.
- Measurement error has its greatest effect on shorter pairwise distances.
- The influence of lens distortion and perspective error affects pairwise distances in a pattern reflecting the 3D geometry of the face:
    - Pairwise distances involving the ear landmarks are most susceptible to error as they are farthest from the camera focal point at the pronasale (prn) in both x and z (width and depth) dimensions, and form the greatest pairwise distances.
    - Pairwise distances between landmarks on the facial midline are the least affected as they are closest to the camera focal point at the pronasale (prn) and have little relative depth (z dimensionality).
    - Pairwise distances between adjacent landmarks on the nose, eyes, and mouth are susceptible to the greatest influence of measurement error.
- Measurement error is greater at lower pixel resolutions.

The approach taken in this investigation (visualization modeling) offers a structure within which the effects of lens distortion, perspective error, and measurement error on anthropometric facial measurement may be simulated for different lens settings and camera distances. It may be used to estimate the camera distance at which lens distortion and perspective error will become negligible in proportion to measurement error; this appears to be greater than 2.0 m for a 50 mm lens. The approach also offers a means via which the optimum camera focal point could be identified in order to minimize the affects of lens distortion and perspective error, after which existing algorithms for the correction of lens distortion could then be applied (see, for example, Farid and Popescu 2001).

References

Aung, S. C., R. C. K. Ngim, and S. T. Lee, 1995. Evaluation of the laser scanner as a surface measuring tool and its accuracy compared with direct facial anthropometric measurements. *Br. J. Plast. Surg.* 48(8): 527–621.

Farid, H., and A. C. Popescu, 2001. Blind removal of lens distortion. *J. Opt. Soc. Am.* A 18: 2072–78.

Ray, S. F. 2002. *Applied photographic optics*, 2nd ed. St. Louis: Focal Press.

Estimation of Landmark Position Using an Active Shape Model

8

MATTHEW I.S. MAYLIN, ALVARO PALLARES BEJARANO, AND CHRISTOPHER J. SOLOMON

Contents

8.1 Introduction

Manual placement of landmarks is a time-consuming process susceptible to inter- and intraobserver error (see Chapter 3).

In this chapter, a method for the automated placement of a set of anthropometric landmarks onto a dataset of 3D face scans using a 3D active shape model (ASM) is investigated. The potential benefit of this method is that it would allow automated batch placement of landmarks, removing the need for laborious manual intervention. It would also allow landmark placement to be more consistent relative to landmark placement by human operators. Automation would be particularly valuable if it could be applied to landmark placement in a large sample database.

Active shape models have been used the registration and analysis of a variety of different objects (Cootes and Taylor 2001, Cootes and Taylor 2000, Cootes et al. 2001, Cootes et al. 1992, Hill et al. 1996). This analysis uses a morphable 3D shape model of the face similar to that proposed by Blanz and Vetter (1999), and explores automated registration similar to that proposed by Hutton et al. (2001) and Dornaika and Ahlberg (2004). This method differs from previous work, however, in that it attempts to identify a set of facial anthropometric landmarks in 3D using two alternative approaches. The first is the use of a genetic algorithm to evolve face shapes using randomly generated populations to obtain the best model for an unregistered face.

The second uses a gradient based direct search method to similarly search the parametric space during principal components analysis (PCA).

8.2 Method

A random sample of 3D images acquired from 100 subjects with the Geometrix FaceVision® FV802 Series Biometric Camera (ALIVE Tech, Cumming, GA) was used in the investigation. MATLAB® (The Mathworks, Natick, MA) was used to perform mathematical and statistical processing.

The dense point model (DPM) of coordinates representing each face consists of a dense but random distribution of discretely sampled points that model the continuous nature of facial surface. In order to perform a rigorous comparison between faces, landmarks were used to ensure correspondence in position of key points between faces. Ideal landmarks offer exact visual or mathematical criteria for their placement. For example, the corner of an eye has clearly defined visual or mathematical criteria. By contrast, the middle of the cheek is less satisfactory—the smooth nature of this region results in a higher uncertainty in its placement.

In order to build an active shape model (ASM), a statistical shape model was first constructed by manually landmarking a small sample of faces at 36 landmark sites that gave general coverage of the face (Farkas 1994). A subset of 40 faces was landmarked in this way. Figure 8.1 shows an example, with the 36 landmarks placed by projection onto the corresponding 3D face mesh. Cubic interpolation was used to alias between points of the discreetly sampled DPM.

Further mathematical landmarks were generated by interpolation, to produce a smoother mesh of highly correspondent landmarks. This was achieved by calculating the mean of the Procrustes-aligned landmarks. Unique triangles and their midpoints in the *xy* plane were identified by Delaunay triangulation. The *z* plane was then resampled, providing additional landmarks. This process was repeated several times to produce a finer mesh, as illustrated in Figures 8.2 to 8.5.

Procrustes analysis was then completed on the 3D landmarks to remove variations in translation δx, δy, δz, scaling (S), and rotation (Θ), so that all of the landmarks were in a common coordinate frame. These landmarks were then represented by a vector, x.

Principal components analysis (PCA) decomposition was performed on the data to allow the shape to be expressed in terms of a weighted combination of the principal components—the basis functions. Therefore, any configuration of anthropometric landmarks for a face that is well represented by the model population can be approximated by Equation 8.1.

$$x = \bar{x} + \mathrm{Pb} \tag{8.1}$$

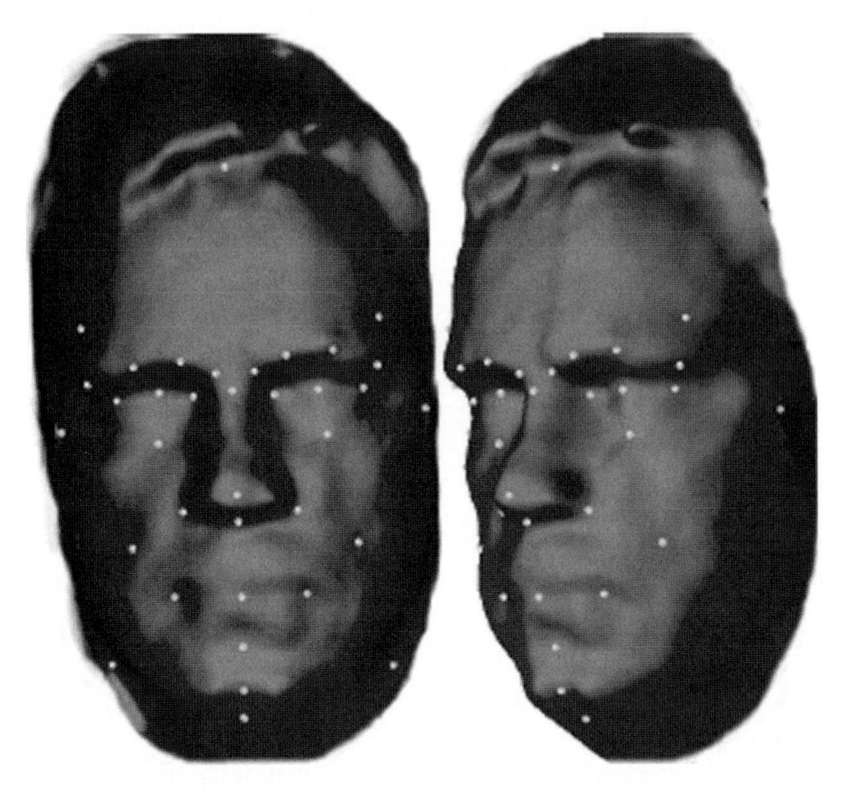

Figure 8.1 Example of an unregistered 3D face surface with manually placed anthropometric landmark points.

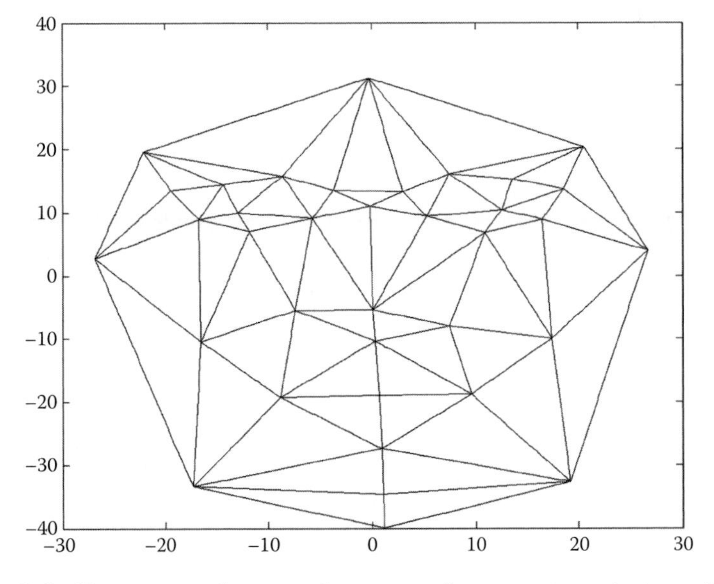

Figure 8.2 Illustration of triangular interpolation: triangulation of initial landmarks.

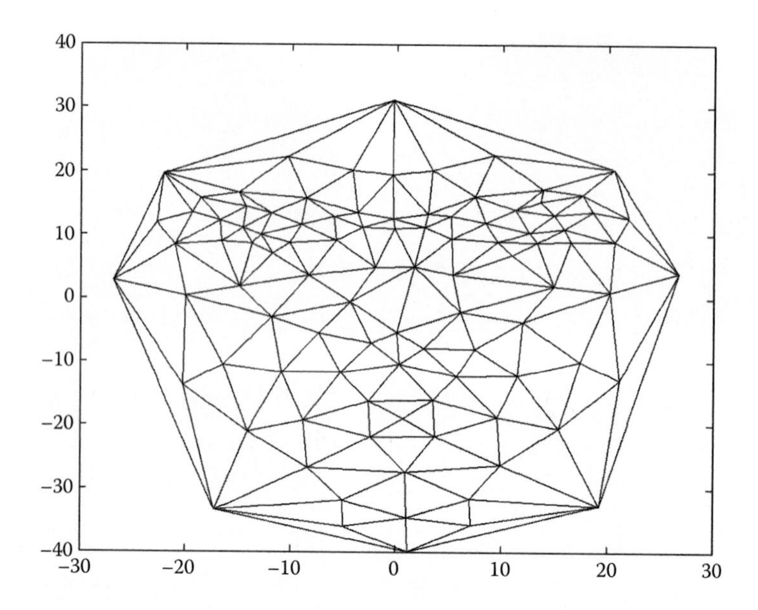

Figure 8.3 Illustration of triangular interpolation: first triangular interpolation.

Where $\bar{x}$ is the mean shape, P is a set of orthogonal modes of variation, and b is a set of shape parameters.

The utility of an ASM approach in automation of landmark placement was then assessed in a subsample of unregistered and unlandmarked faces.

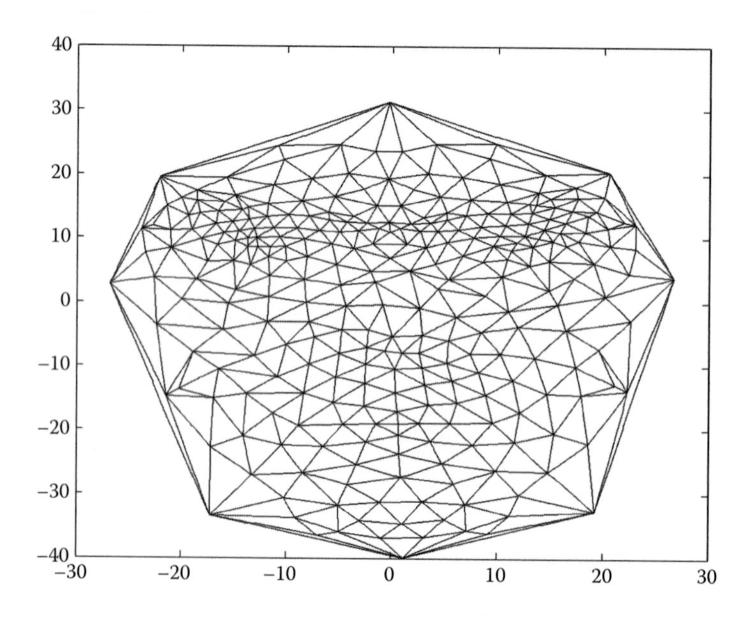

Figure 8.4 Illustration of triangular interpolation: second triangular interpolation.

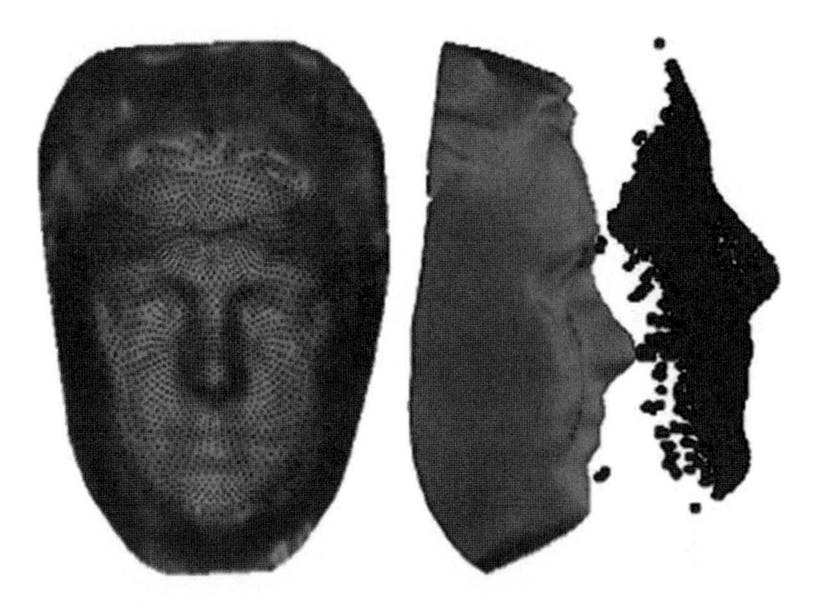

Figure 8.5 Unregistered face displayed as a surface compared with the estimated face displayed as points in anterior (left) and right lateral (right) views.

Each new unregistered face x^{unreg} was first approximated to the model x^{est}, given in Equation 8.2, by identifying optimal parameters of scaling S, rotation Θ, translation $\delta_{x,y,z}$, and shape parameters b, so that the mean-squared difference between x^{unreg} and x^{est} was minimized.

$$x^{est} = \Theta\ \{S[\ \bar{x}\ + \mathrm{Pb}]\} + \delta_{x,y,z} \tag{8.2}$$

In order to standardize x^{unreg} and x^{est}, the z-values of the estimated model (x', y') were interpolated to the x, y coordinates (x, y), producing $z^{unreg}\ (x, y)$, as shown in Equation 8.3.

$$z^{est}(x', y') \xrightarrow{x,\, y\ interpolation} z^{est}(x, y) \tag{8.3}$$

With z^{unreg} and z^{est} in the same coordinate system, the error metric ξ was used to determine the goodness of fit. This was the sum of the square difference of the estimated $z^{unreg}\ (x, y)$ and unregistered $z^{est}\ (x', y)$, described by Equation 8.4.

$$\xi = \frac{1}{n} \sum_{n} \left\{ z^{unreg}(x, y) - z^{est}(x, y) \right\}^2 \tag{8.4}$$

To minimize ξ in Equation 8.4, a set of optimal parameter values for the variables b, δx, δy, δz, S, and Θ were sought using genetic algorithm (GA) and gradient-based (GB) approaches in MATLAB. The algorithms were initialized by making a simple approximate alignment of the basic shapes of the estimated and unregistered meshes—for example, at the pronasale (tip of the nose). The scaling was set to 1, initially, and the mean face shape $\bar{x}$ was used, with parameter b set to 0. Several variations were investigated in order to identify the best performing algorithm of each type. These were then assessed for landmark placement accuracy in an out-of-sample dataset of 30 3D facial images by comparing landmark position with those of a control face manually landmarked 20 times.

8.3 Results

The mean subtracted variance of Euclidean distances was used as the measure of error in manual landmarking. Assuming the mean of these placements to be an estimate of the true position, the measured deviation over 20 trials offered a measure of the inaccuracies of manual placement. Results for GA, GB, and manual landmark placement are shown in Table 8.1, and Figures 8.6 and 8.7.

Table 8.1 and Figure 8.6 show that both ASM methods are sources of considerably higher error relative to the manual method, and that the ASM methods generate errors covering a greater range. The distribution of the manually placed landmark errors is more highly clustered toward lower landmark placement error, and has a lower mean and standard deviation as shown in Table 8.1.

Figure 8.7 shows that the GB method generally had a slightly lower landmark placement error than the GA method, also reflected in the mean and standard deviation of its placement errors shown in Table 8.1.

Table 8.1 Statistical Distributions of Landmarking Errors Associated with GA, GB, and Manual Landmarking, Displaying the Mean and Standard Deviation (SD) for each Method

Method	Mean (mm)	SD (mm)
GA	12.630	12.279
GB	9.200	9.456
Manual	1.008	2.323

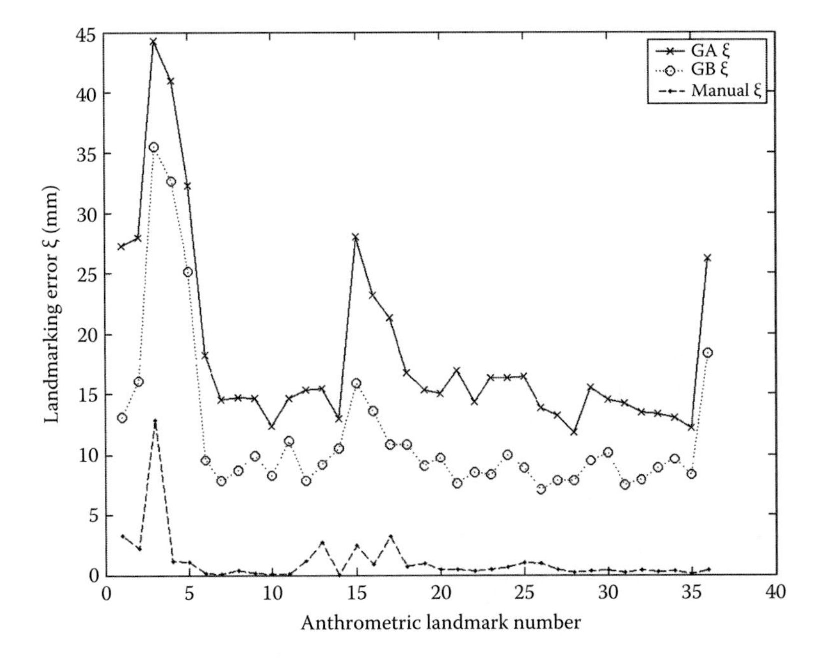

Figure 8.6 Chart showing landmarking errors for 36 facial anthropometric landmarks placed using the GA, GB, and manual methods.

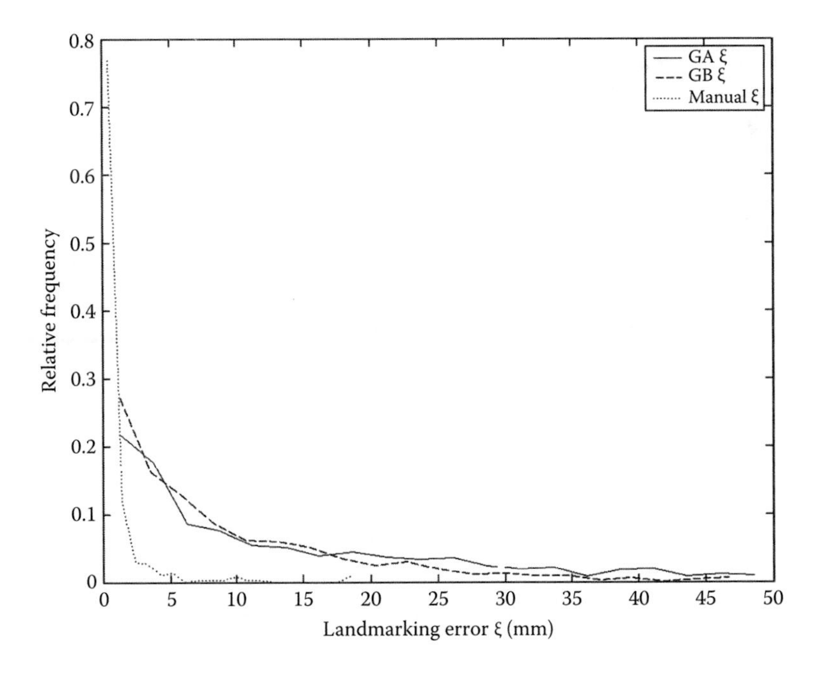

Figure 8.7 Relative frequencies of landmarking errors for the two optimization techniques.

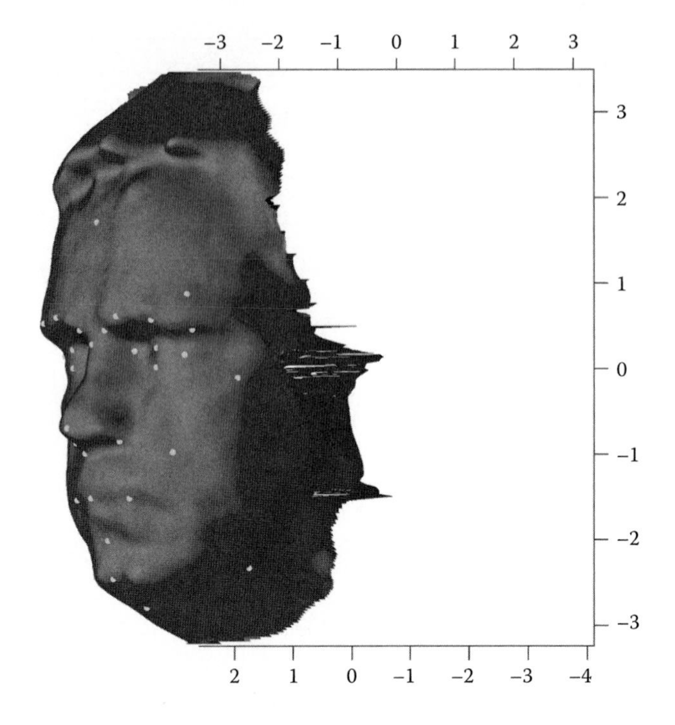

Figure 8.8 Visualization of automatically placed landmarks.

Figure 8.8 shows a visualization of the landmarks placed automatically (compare with Figure 8.1).

Higher errors are associated with landmarks at the periphery of the face. This suggests that the ASM model or optimization method is less effective at accurately modeling and placing these landmarks. It may be anticipated that, as these landmarks are all located around the edge of the face model, triangular interpolation would cause the subsequent density of the face mesh to be lower in these regions. This would cause the model to be less constrained in these regions, resulting in larger errors in the associated landmarks.

8.4 Discussion

The large mean errors—of 9.2 and 12.6 mm—encountered in this investigation indicate that reliable estimation of landmark position using ASM is a nontrivial problem. Furthermore, combined error would, at present, probably exceed differences due to face shape variation, making such measurements ineffectual as a tool in facial comparison.

Optimization methods have a significant influence on performance and another approach may be more successful. While attempts were made to reduce the convergence to false optimizations—by initial conditioning and constraints—these were not very successful. This is more apparent in the GA optimization method, where its inherent ability to search a greater proportion of the parametric space leads to frequent arrival at false optimizations and a larger associated error. Possible causes for the performance of these methods may be that this system has multiple optimal solutions or various local minima. It may be possible to improve performance by use of an appearance model that combines knowledge of both shape and texture, and landmark texture, in regions with high textural detail.

References

Blanz, V., and T. Vetter, 1999. A morphable model for the synthesis of 3D faces. *SIGGRAPH'99 Conf. Proc.* 187–194.

Cootes, T. F., and C. J. Taylor, 2001. Statistical models of appearance for medical image analysis and computer vision. *Proc. Int. Soc. Opt. Eng.* 4322: 236–48.

Cootes, T. F., and C. J. Taylor, 2000. Combining elastic and statistical models of appearance variation. *Proc. Eur. Conf. Comput. Vision* 2000 1: 149–63.

Cootes, T. F., G. J. Edwards, and C. J. Taylor, 2001. Active appearance models. *IEEE Trans. Pattern Anal. Machine Intelligence* 23(6): 681–85.

Cootes, T.F., C. J. Taylor, D. H. Cooper, and J. Graham, 1992. Training models of shape from sets of examples. *Proc. Br. Machine Vision Conf.* 1992: 9–18.

Dornaika, F., and J. Ahlberg, 2004. Fast and reliable active appearance model search for 3-D face tracking. *IEEE Trans. Syst. Man Cybernetics* Part B, 34(4): 1838–53.

Farkas, L. G., 1994. *Anthropometry of the head and face*, 2nd ed. New York: Raven Press.

Hill, A., T. F. Cootes, and C. J. Taylor, 1996. Active shape models and the shape approximation problem. *Image Vision Comput.* 14(8): 601–608.

Hutton, T. J., B. F. Buxton, and P. Hammond, 2001. Dense surface point distribution models of the human face. *Proc. IEEE Workshop on Mathematical Methods in Biomedical Image Analysis* 2001: 153–60.

Generation of Values for Missing Data

LUCY MORECROFT AND NICK R.J. FIELLER

Contents

9.1 Introduction

Procrustes analysis and other statistical methods require that there be no missing values in the dataset. This chapter summarizes an investigation of incomplete data handling using the EM algorithm.

It is important to distinguish between the different types of incomplete data. Incomplete values can either be missing completely at random or informatively missing. Here, the data can be regarded as informatively missing data: the variables that were missing in the landmark configurations followed similar patterns, with many missing values around the ears and areas where hair or facial hair obstruct the position of the landmark points.

9.2 The Data

The facial landmark database consists of 3D landmark configurations of 30 different facial landmark points. The data were taken from 2966 different facial images. The landmarks were manually placed on the images by one of six different operators; the process involved positioning and clicking with the mouse. Each facial image had landmarks placed on it twice, once each by different operators, giving a total of 5932 landmark configurations.

For many facial images the full set of 30 landmarks was not able to be collected. Certain facial attributes obstruct the view of some landmark points, particularly around the mouth and ears, where beards and hair can mask the true position of landmarks. When this occurred in a facial image the landmark placement operators were instructed not to place the landmark points unless they were clearly distinguishable. So, the resulting facial landmark database of 5932 configurations is incomplete, that is, not all configurations contain values for all 30 landmark points.

An investigation into the data showed that 3254 configurations had complete values for all 30 landmark points, which was just over half of the data collected. Table 9.1 lists the 30 landmarks and the percentage of incomplete data for each one.

Table 9.1 Percentage of Incomplete Data for Each of the 30 Landmark Points

#	Landmark	Label	Complete	Incomplete	% Incomplete
1	Glabella	g	5639	294	4.96
2	Sublabiale	sl	5565	368	6.20
3	Pogonion	pg	5693	240	4.05
4	Endocanthion left	en l	5915	18	0.30
5	Endocanthion right	en r	5915	18	0.30
6	Exocanthion left	ex l	5915	18	0.30
7	Exocanthion right	ex r	5911	22	0.37
8	Center point of pupil left	p l	5906	27	0.46
9	Center point of pupil right	p r	5903	30	0.51
10	Palpebrale inferius left	pi l	5921	12	0.20
11	Palpebrale inferius right	pi r	5914	19	0.32
12	Subnasion	se	5717	216	3.64
13	Alare left	al l	5931	2	0.03
14	Pronasale	prn	5914	19	0.32
15	Alare right	al r	5925	8	0.13
16	Highest point of columella prime left	c′ l	5909	24	0.40
17	Highest point of columella prime right	c′ r	5902	31	0.52
18	Labiale superius	ls	5883	50	0.84
19	Labiale inferius	li	5911	22	0.37
20	Stomion	sto	5594	339	5.71
21	Cheilion left	ch l	5867	66	1.11
22	Cheilion right	ch r	5867	66	1.11
23	Superaurale left	sa l	4710	1223	20.61
24	Superaurale right	sa r	4696	1237	20.85
25	Subaurale left	sba l	5440	493	8.31
26	Subaurale right	sba r	5448	485	8.17
27	Postaurale left	pa l	4948	985	16.60
28	Postaurale right	pa r	4889	1044	17.60
29	Otobasion inferius left	obi l	5504	429	7.23
30	Otobasion inferius right	obi r	5463	470	7.92

Table 9.2 Incomplete Cases by Number of Missing Landmarks

Number of Missing Landmarks in Case	Number of Cases
1	809
2	670
3	335
4	309
5	152
6	129
7	54
8	140
9	36
10	26
11	7
12	4
13	2
14	2
16	2
17	2
27	1

The landmarks with by far the highest percentage of incomplete data were the postaurale (left and right) and the superaurale (left and right); these two points are located on the back and top of the ear respectively, and so both will be frequently obstructed by hair.

A high number of landmarks have less than 1% of their data incomplete, it was these landmarks that were taken and used as a core set of data to use with the EM algorithm to gain completeness throughout the whole data set.

Table 9.2 illustrates the number of cases that had incomplete data, and the number of landmarks that were missing. One configuration had three complete landmarks and 27 incomplete; this observation was discarded, as there was not enough data there to accurately execute the EM algorithm.

9.3 The EM Algorithm for Incomplete Data Handling

Many techniques for the exact analysis of incomplete data appear to start by estimating the missing values. It is important to appreciate that this is just a numerical device in an iteration system.

The most widely used iterative procedure for handling incomplete data is the estimate and modify (EM) algorithm. This method was originally proposed for much more complex statistical analyses than handling the analysis of incomplete data.

The algorithm cycles between estimating the missing value and then using this to modify the statistical model. With each iteration, the modified model is used to provide the next estimate of the missing value. This procedure can be shown to converge to precisely the exact analysis (that which would be obtained by solving the set of implicit equations for the unknown parameters of the multivariate normal statistical model) and to do so reasonably efficiently. Thus, the EM algorithm provides a convenient and intuitive method for the iterative solution of a set of implicit equations. With the usual models based on multivariate normal distributions, both the E and the M steps are given by the solutions to explicit equations that can be provided in closed form, and are easy and quick to calculate.

9.4 The Model

The real complete data—3254 landmark configurations of 30 landmarks—was modeled using a multivariate normal distribution. To check the fit of the model the Mahalanobis distance of each configuration to the mean was plotted against the appropriate Chi-squared distribution. Figure 9.1 displays this plot for the real complete landmark data.

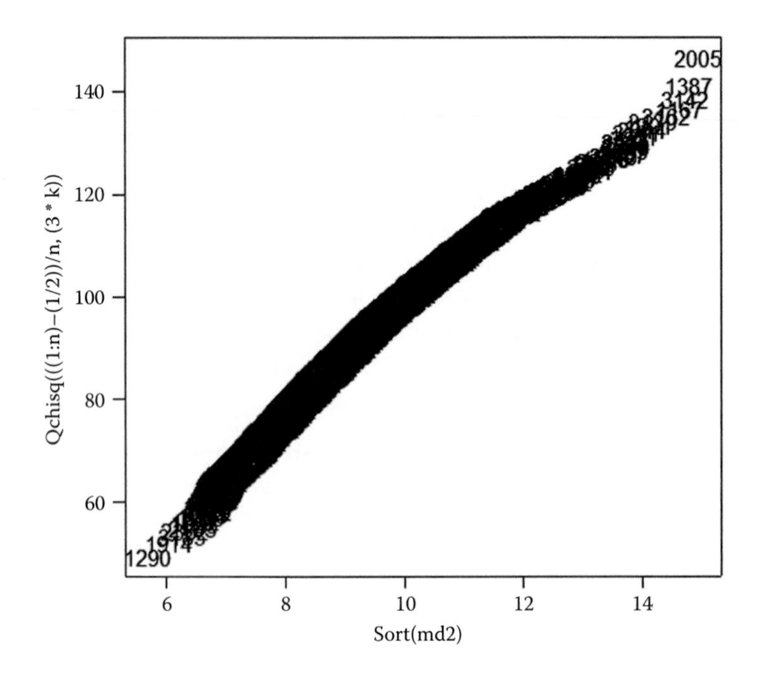

Figure 9.1 Plot demonstrating multivariate normality of real complete data in 3254 complete configurations of 30 landmark points.

The model is a good fit for the data, indicated by the linearity of the plot. It was this model for the real complete data that was used with the EM algorithm to generate values to make the whole dataset complete, yet still maintain the same model.

9.5 Analysis and Results

For the main part of the analysis 17 landmarks were taken as the core set (see Table 9.3), these were complete for the majority of the data set, having 5735 complete configurations.

The 5735 cases with complete observations were used with the EM algorithm to generate complete data for the other 13 landmarks under study. This process was carried out by taking one landmark at a time. Landmark 1 was addressed first, and all configurations within the 5735 that had landmark 1 incomplete—but the entire core landmarks complete—were extracted and used to perform the EM algorithm. The em.norm function in R was used to carry out the procedure. The other incomplete landmarks were addressed in the same way until all 30 landmarks were complete for the 5735 configurations.

Table 9.3 Core set of 17 Landmarks Used Initially with the EM Algorithm

Landmark Number	Landmark	Label
4	Endocanthion left	en l
5	Endocanthion right	en r
6	Exocanthion left	ex l
7	Exocanthion right	ex r
8	Center point of pupil left	p l
9	Center point of pupil right	p r
10	Palpebrale inferius left	pi l
11	Palpebrale inferius right	pi r
13	Alare left	al l
14	Pronasale	prn
15	Alare right	al r
16	Highest point of columella prime left	c′ l
17	Highest point of columella prime right	c′ r
18	Labiale superius	ls
19	Labiale inferius	li
21	Cheilion left	ch l
22	Cheilion right	ch r

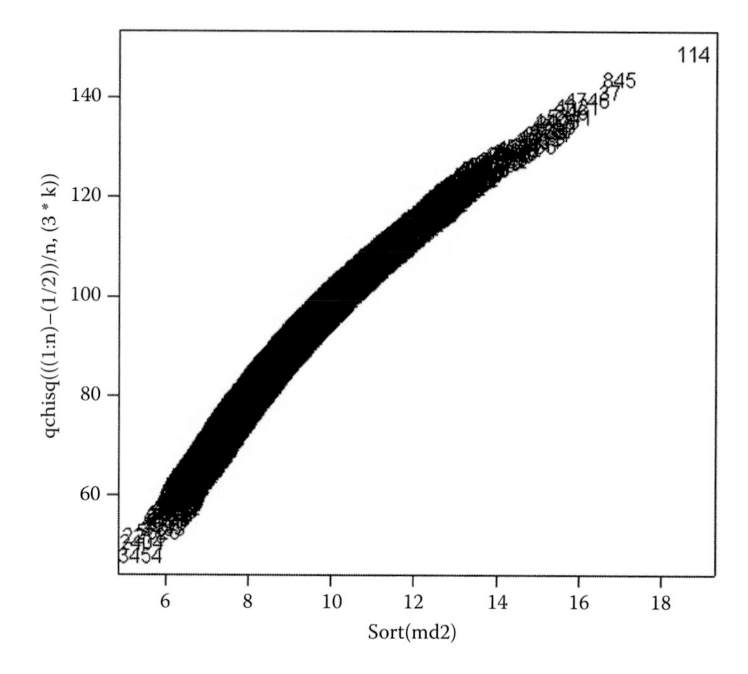

Figure 9.2 Plot demonstrating multivariate normality of EM complete data in 5735 configurations of 30 landmark points, made complete by using 17 core landmarks (see Table 9.3) with the EM algorithm.

The imputation of each separate landmark was based only on real complete configurations, and no previous imputed value was used in the calculations.

To check the model of the EM completed data for 5735 configurations, the Mahalanobis distance of each configuration to the mean was again plotted against the appropriate Chi-squared distribution (see Figure 9.2), and the model fit is very good.

There were 199 incomplete configurations in the database that could not be dealt with by using the core subset of landmarks, because these configurations had incomplete observations in the core set. To complete these configurations, the same technique was applied using a different core subset of eight landmarks (see Table 9.4).

Of the 199 remaining incomplete configurations, 135 had complete values for the new core set of eight landmarks. In the same way as described above, one incomplete landmark was taken at a time and used—with the complete landmarks from the core set—to carry out the EM algorithm, until the 135 configurations were all complete for all 30 landmark points. These configurations were bound to the initial 5735 EM-completed configurations to make a total of 5870 complete configurations. The model was again checked for correspondence to a multivariate normality by plotting Mahalanobis distances against the appropriate Chi-squared distribution (see Figure 9.3).

Table 9.4 Core Set of Eight Landmarks Used in the Second EM Algorithm

#	Landmark	Label
4	Endocanthion left	en l
5	Endocanthion right	en r
6	Exocanthion left	ex l
10	Palpebrale inferius left	pi l
11	Palpebrale inferius right	pi r
13	Alare left	al l
14	Pronasale	prn
15	Alare right	al r

Figure 9.3 shows slightly more curvature than the previous two models shown in Figures 9.1 and 9.2, but the model is still a good fit. The curvature is probably due to the fact that the second EM analysis was carried out by using only a few complete landmarks to estimate a high number of incomplete ones. This is finding is evident again in Figure 9.4, where the EM algorithm was applied to a

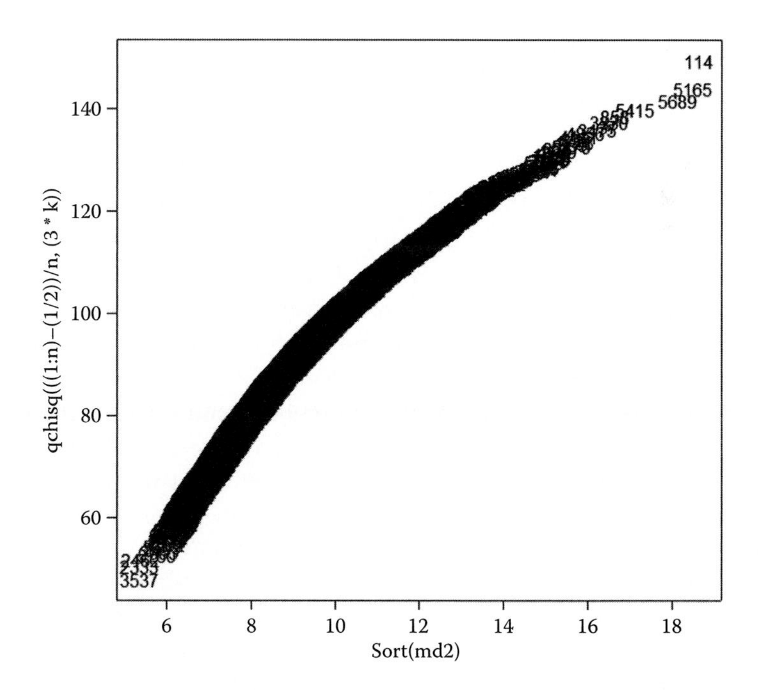

Figure 9.3 Plot demonstrating multivariate normality of EM complete data in 5870 configurations of 30 landmark points. Data were made complete by first using 17 core landmarks (Table 9.3) with the EM algorithm to estimate the first 5735 observations, then eight core landmarks (Table 9.4) to estimate the remaining 135 observations.

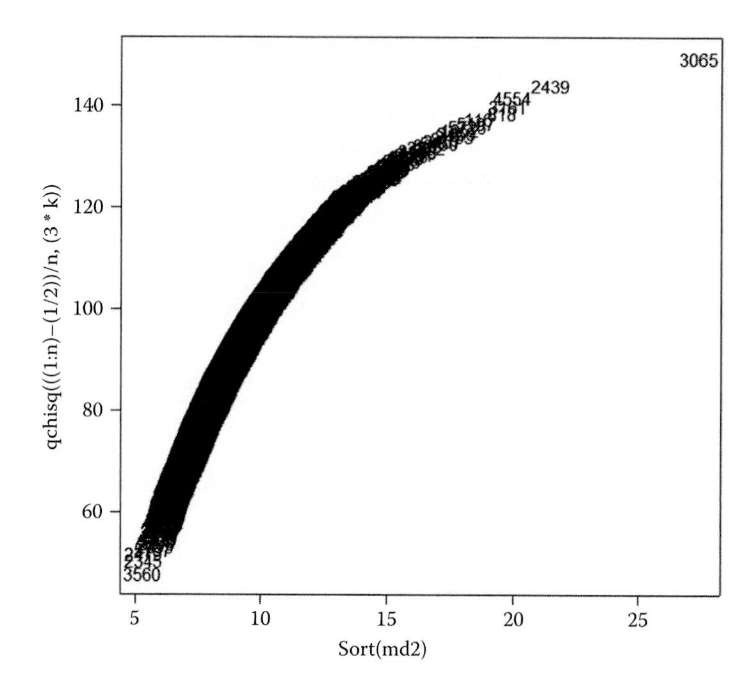

Figure 9.4 Plot demonstrating multivariate normality of EM complete data in 5907 configurations of 30 landmark points. Twelve core landmarks were used to estimate the 37 configurations in addition to the ones in Figure 9.3.

further subset of core landmarks to try to estimate the remaining 64 incomplete configurations that could not be completed in the first two applications.

This third set of core landmarks only completed 37 of the remaining 64 incomplete configurations. When compared to Figure 9.1—the model for the real complete data—the plot shown in Figure 9.4 implies that the model is not a good fit to the data: the linearity of the plot is not good. The 64 remaining incomplete configurations had high numbers of incomplete variables even after the first two applications of the EM algorithm. It is apparent that there was not enough real data present to gain good estimates for the incomplete data. The final 64 observations were, therefore, excluded from the 5870 configurations in the final EM-complete landmark database.

9.6 Summary

The EM algorithm was used to investigate completing 2678 cases in the landmark database of 5932 records where landmark data was missing. The EM algorithm was applied three times. The first two applications—using core datasets in which 17 and 8 landmarks were complete, respectively—successfully completed missing landmarks in correspondence with a multivariate normal model. These were retained in the EM-complete landmark database of 5870 configurations. A final application of the EM algorithm using a core dataset with 12 complete landmarks showed poor correspondence to a multivariate normal model and was excluded.

Admissibility

10

XANTHÉ MALLETT

Contents

10.1 Introduction

This chapter is a brief exploration of general issues of courtroom admissibility, including the definition of an expert witness and their role in court. This is a rapidly developing and often controversial field, and in order to introduce the reader to debate surrounding key cases and issues in forensic human identification, references have been chosen that are easily accessible on the Internet.

10.2 The Expert, the Method, and the Evidence

Exactly what criteria require fulfillment by an individual prior to their being accepted as an expert, and what they can offer evidentially in an individual case, is a hotly debated topic. The contention primarily focuses on the qualifications required in order to be accepted as an expert, although ultimately it is for the judge to decide who can give evidence and who cannot. In judicial proceedings, a court should only be interested in specific, rather than general, knowledge that an expert offers as opinion, and experts should only be used to clarify necessary information for the jury. The degree of certainty that is afforded to specific evidence is particularly relevant when a new technique is being developed, and it must be sufficiently evaluated and tested. To ensure the evidence is not declared inadmissible, not only must the expert be accepted but the evidence must also be independently scrutinized. Subsequently, the level of authority with which an opinion is formed and offered must be such that the court can readily recognize the relevance of the opinion.

To be admissible, a new method needs to be validated and its practitioners need to be recognized as experts. This requires a number of specific criteria to be fulfilled.

Three key cases require consideration with regard to the admissibility of expert testimony: *Daubert v. Merrell Dow Pharmaceuticals, Inc., 509 U.S. 579 (1993)*; *General Electric Co. v. Joiner, 522 U.S. 136 (1997)*; and *Kumho Tire Co., Ltd., v. Carmichael, 526 U.S. 137 (1999).*

10.2.1 *Daubert v. Merrell Dow Pharmaceuticals*

On June 28, 1993, the U.S. Supreme Court issued an opinion that laid out specific guidelines for determining what scientific evidence is admissible in court. These guiding principles were established in *Daubert v. Merrell Dow Pharmaceuticals, Inc., 509 U.S. 579 (1993).* This tort lawsuit held that scientific evidence must be subjected to reliability testing, and that judges must act as "gatekeepers," determining if evidence is reliable (Gebauer 2004, Randerson and Coghlan 2004, Shaw 2001, Westfall 2004). The court also established parameters by which reliability can be judged. This list is not definitive, however, and the ruling is clear in its recommendations that judges must be open to employ criteria of their own (Hileman 2003). These parameters, known as the "Daubert factors" comprise the following:

- The evidence must be based on a testable theory or technique.
- The expert's theory or technique can be, or has been, tested.
- The theory or technique has been subject to peer review.
- There exists a known or potential associated error rate of the theory or technique when applied; there are controls and standards in existence.
- The theory or technique has been generally accepted in the relevant scientific community.

10.2.2 *General Electric Co. v. Joiner*

In 1997 the trial court's gatekeeper role was broadened during the hearing of *Electric Co. v. Joiner, 522 U.S. 136 (1997)*, when it was called upon to assess all testimony that can be described as scientific, technical, or of other specialized knowledge. *Joiner* concluded that it was the job of the district courts to scrutinize the expert's general method as well as the reliability of his or her reasoning, and therefore the expert's analysis; the evidence can be declared inadmissible if the court concludes that there is too large an analytical gap between the data and the opinion proffered.

10.2.3 *Kumho Tire Co., Ltd. v. Carmichael*

As a result of the *Daubert* and *Joiner* rulings, a number of questions arose regarding what evidence was affected by the Daubert standard. The Supreme Court settled these in *Kumho Tire Co., Ltd., v. Carmichael, 526 U.S. 137 (1999)*, a typical product case which stipulated that the trial judge's gatekeeper role should be exercised in relation to all expert witness testimony. Furthermore, it stated that technical and all other expert evidence should be both relevant and based upon a reliable foundation prior to admission. It established that a trial judge has the authority to determine whether and to what extent the enumerated Daubert factors are applicable to a case (Hileman 2003).

The Supreme Court's "expert evidence trilogy" tightened the rules for the admissibility of expert evidence (Aptel 2004, Gebauer 2004, Hileman 2003), and consequently the admission of potentially damaging and inappropriate expert evidence was reduced (Bernstein 2002).

Any benefits the triad of rulings may have had was limited, however, as not all U.S. states adopted Daubert. The solution to the apparent recalcitrance of state courts to accept the appropriate strict standards for the admissibility of expert testimony was for state legislatures to implement an amended version of Federal Rule 702, the original of which was far more ambiguous (Bernstein 2002).

10.2.3.1 Federal Rule of Evidence 702

Effective as of December 1, 2000, the new Federal Rule of Evidence 702 incorporated the expert reliability requirements announced by the Supreme Court in *Daubert*, although the Daubert factors are not specified in detail. Instead, the holdings of *Daubert*, *Joiner*, and *Carmichael* are subsumed within the rule's three-pronged requirements which decree that an expert may testify if (Gebauer 2004):

- The testimony is based upon sufficient data or facts.
- The testimony is the product of reliable principles and methods.
- The witness has applied the methods and principles reliably to the facts of the case.

In addition, Federal Rule 702 requires that the trial court considers the reliability of the expert's opinion in its entirety, based on this triad of requirements. Consequently, any step that renders the testimony unreliable, renders the evidence inadmissible (Gebauer 2004). The standard of evidential reliability enforced by Federal Rule of Evidence 702 requires that expert testimony be judged by real-world, objective measures. Therefore, the trial judge should decide which expert opinions are admitted in the courtroom based upon reliability standards developed, for the most part, outside of the courtroom (Gebauer 2004). The duties and responsibilities of expert witnesses in civil cases thereby include the following (see *National Justice Compania Naviera SA v. Prudential Life Assurance Co. Ltd. (No. 1) [1995] 1 Lloyd's Rep 445).*

An expert witness should:

- Provide independent assistance to the court by way of objective unbiased opinion in relation to matters within their expertise, and should never assume the role of advocate.
- State the facts or assumptions on which their opinion is based. They should not neglect to consider material facts which detract from their concluded opinion.
- Make it clear when a particular question or issue falls outside their expertise.
- Make it clear to the jury if their opinion is not properly researched because they consider that insufficient data is available, indicating that the opinion is no more than a provisional one.
- Communicate to the other side without delay if, after exchange of reports, an expert witness changes their view on a material matter.

Expert evidence should:

- Be the independent product of the expert uninfluenced as to form or content by the exigencies of litigation.
- Provide to the opposite party at the same time as the exchange of reports any evidence referring to photographs, plans, calculations, survey reports or other similar documents.

10.3 Discussion and Conclusion

Expert evidence should be complete, credible, and thorough. The work of all forensic scientists and the evidence they give in court should be able to forestall or defeat any challenge, as these are an integral part of any justice system.

The key components in assessing the authority of any knowledge presented in court are:

- The authority of the expert
- The authority of the knowledge

Authority of knowledge has its basis in validation, relevance, and currency. The authority of the expert is based on their competence to evaluate the knowledge, and their currency in doing so. Therefore, each expert must be assessed in view to examining the justification, relevance, and currency of each premise in the opinion. Furthermore, the degree to which an expert uses the expertise of others as a premise for their own opinion, their ability to understand and ingest input, and the authority of these other experts upon which they rely are all components which must be considered when assessing the authority of the opinion.

The expert's opinion must be appropriate and comprehensible. Training in the relevant points of law and the presentation of expert evidence is a key requirement. The sources of conclusions, and therefore their opinion, must be explicit.

Finally, continuing research that contributes to the advancement of the technique is required and the research agenda must also enable criminal justice practitioners to understand the uses and limitations of the science.

References

Aptel, D., 2004. *Expert testimony under Daubert*. http://www.birthinjuryinfo.com/biexperts.html (accessed March 30, 2009).

Bernstein, D. E., 2002. Improving the qualifications of experts in medical malpractice cases. *Law Probab. Risk* 1:9–16.

Gebauer, M. E., 2004. *The "What" and the "How" of Daubert challenges to expert testimony under the new Federal Rule of Evidence 702.* http://www.eckertseamans.com/file/pdf/publications/megjuly2001.pdf (accessed March 30, 2009).

Hileman, B., 2003. *Daubert rules challenge courts.* http://www.ourstolenfuture.org/press/2003/2003-0707-CEN-daubert.htm (accessed March 30, 2009).

Randerson, J., and A. Coghlan, 2004. *Investigation: Forensic evidence in the dock.* http://www.newscientist.com/news/news.jsp?id=ns99994611 (accessed February 11, 2004).

Shaw, J., 2001. *Toolmark identification received a (Fyre-Daubert) body blow in Florida.* http://www.forensic-evidence.com/site/ID/toolmark_id.html (accessed March 30, 2009).

Westfall, B., 2004. *Validity of earprint evidence questioned: Judge hears arguments on whether suspect's earprint should be part of trial.* SCAFO (Southern Californian Association of Fingerprint Officers) Online. http://www.scafo.org/library/130204.html (accessed March 30, 2009).

Application Toolset 11

LUCY MORECROFT AND MARTIN PAUL EVISON

Contents

11.1 Introduction

For computer-assisted forensic facial comparison to be of practical value, theoretical models must be tested, and practical tools must be produced. The process of research, development, testing, and training is a minimum requirement for validation and admissibility in court.

This chapter describes prototypic application tools for use in forensic facial comparison and their testing. The tools are based on the research described in the preceding chapters.

The research is based on an anthropometric approach to facial comparison utilizing the traditional landmarks of Farkas (1994) and demonstrates that substantial research databases may be collected in 3D. The research anticipates, however, that questioned images—of offenders, for example—will be encountered in 2D, at least for the foreseeable future. Similarly, the suspect or defendant cannot—realistically—be expected to routinely submit to having his or her facial image captured in 3D. Analog and digital photographs or video stills of the suspect are far more likely to be encountered in forensic facial comparison cases for the time being.

Three application tools are therefore described: a 2D landmarking tool, a 2D-2D comparison tool and a 3D-2D projection tool. Rudimentary testing is also described.

The tools were developed in Visual Basic—as this environment offered the best option for rapid programming and prototype development in Microsoft® Windows using an iterative process of design and end user testing. In addition, there is a facility in Visual Basic through which program calls may be made to routines written in R.

11.2 2D Landmarking Tool

A primary requirement for a 2D landmarking application tool was identified—one capable of precisely placing the *x* and *y* coordinates of selected set of anthropometric landmarks on a 2D facial image. The final prototype offers standard and drop down menu options for file handling; viewing, placing, and removing landmarks; zooming in and out, and Help; and reticules permitting visibility of the underling image surface when placing the landmark in a click-and-drag operation with the mouse. A screenshot illustrating the 2D landmarking tool is shown in Figure 11.1.

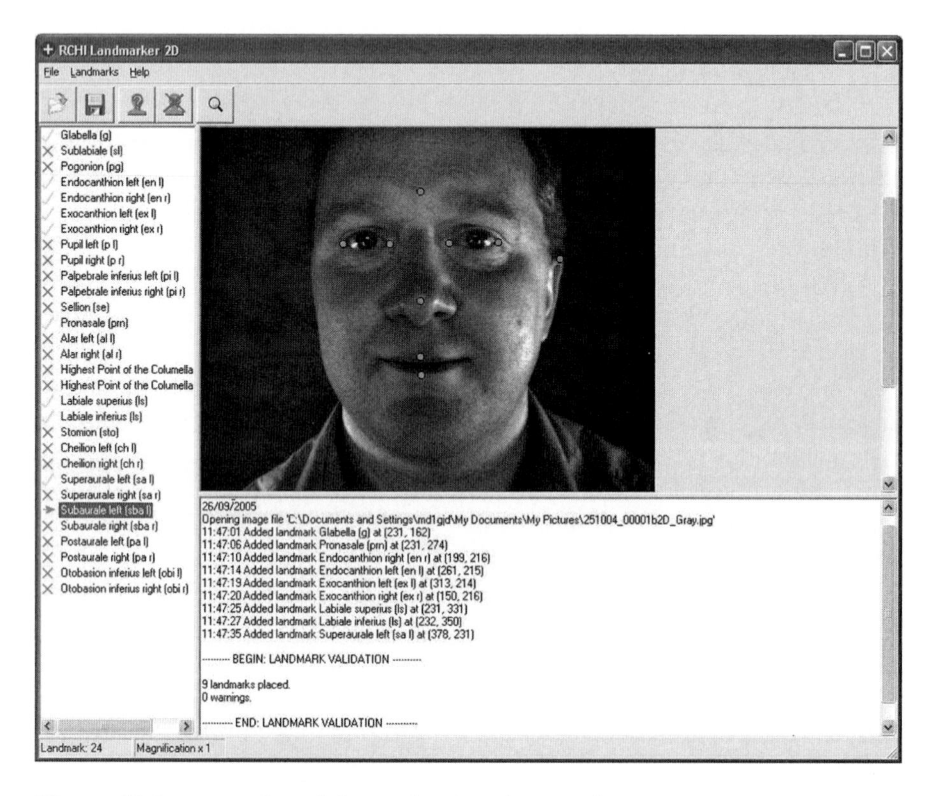

Figure 11.1 Screenshot of the 2D landmarking tool.

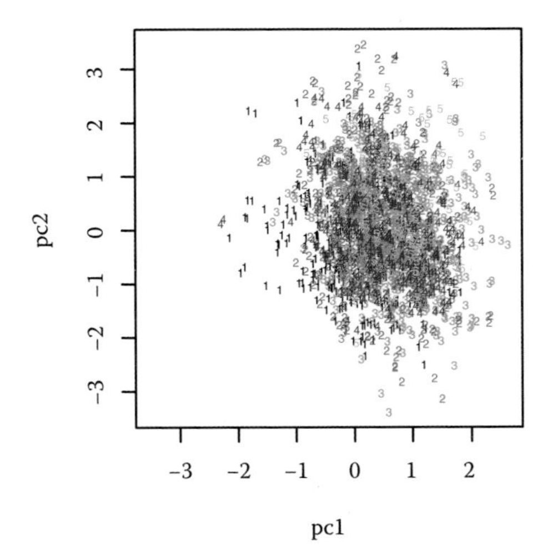

Color Figure 4.6 Plot of PC1 versus PC2 for males by age group (1: under 25; 2: 25–34; 3: 35–44; 4: 45 and over).

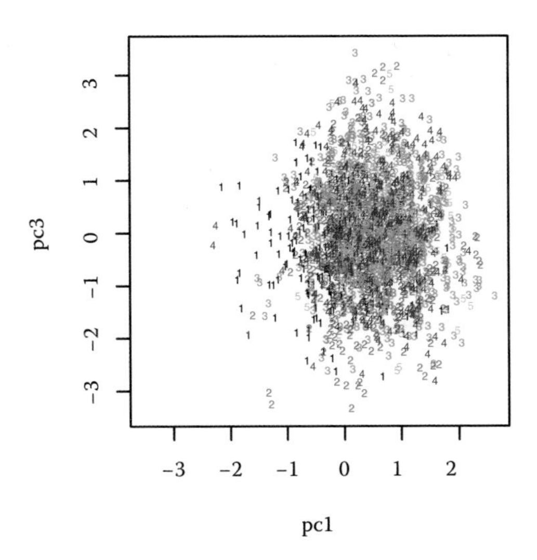

Color Figure 4.7 Plot of PC1 versus PC3 for males by age group (1: under 25; 2: 25–34; 3: 35–44; 4: 45 and over).

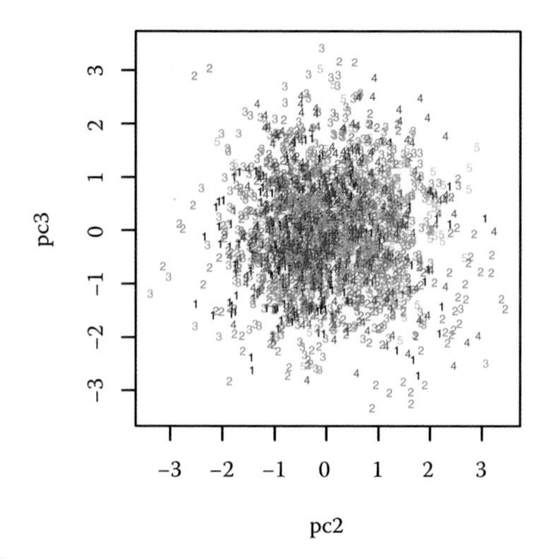

Color Figure 4.8 Plot of PC2 versus PC3 for males by age group (1: under 25; 2: 25–34; 3: 35–44; 4: 45 and over).

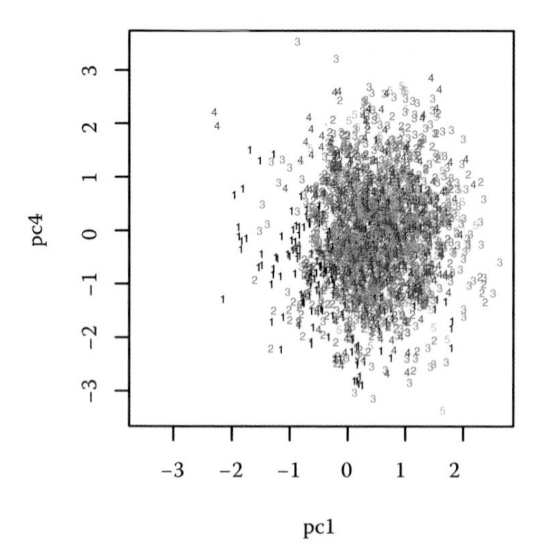

Color Figure 4.9 Plot of PC1 versus PC4 for males by age group (1: under 25; 2: 25–34; 3: 35–44; 4: 45 and over).

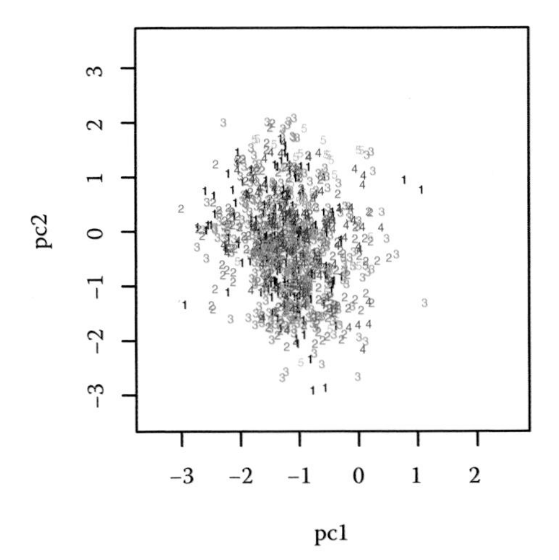

Color Figure 4.10 Plot of PC1 versus PC2 for females by age group (1: under 25; 2: 25–34; 3: 35–44; 4: 45 and over).

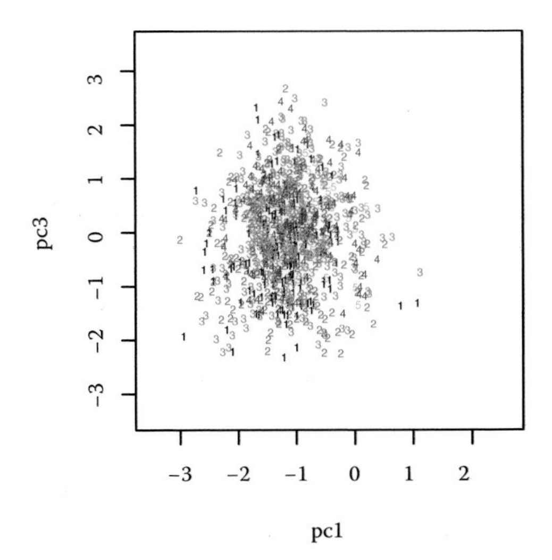

Color Figure 4.11 Plot of PC1 versus PC3 for females by age group (1: under 25; 2: 25–34; 3: 35–44; 4: 45 and over).

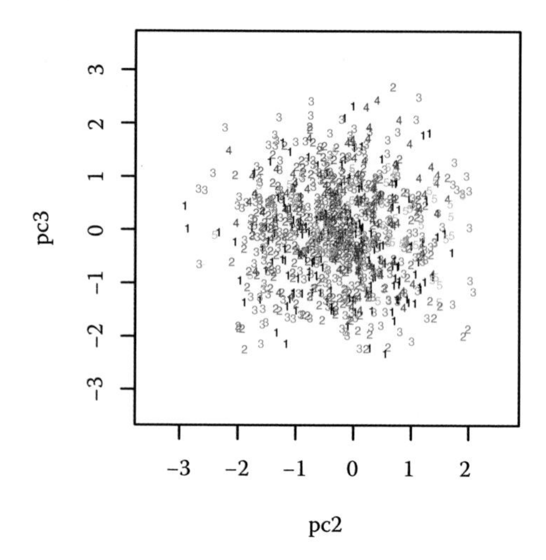

Color Figure 4.12 Plot of PC2 versus PC3 for females by age group (1: under 25; 2: 25–34; 3: 35–44; 4: 45 and over).

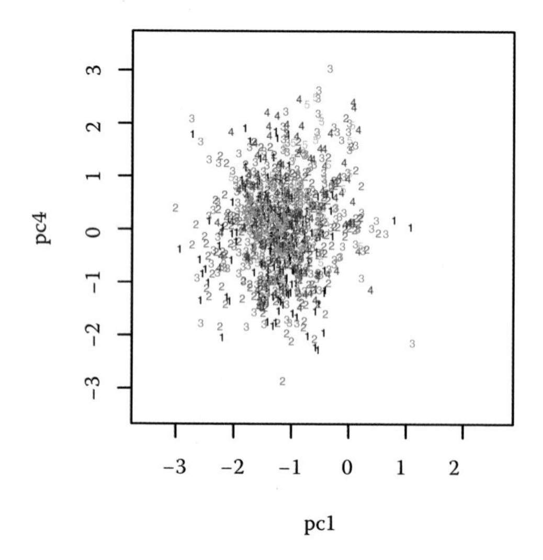

Color Figure 4.13 Plot of PC1 versus PC4 for females by age group (1: under 25; 2: 25–34; 3: 35–44; 4: 45 and over).

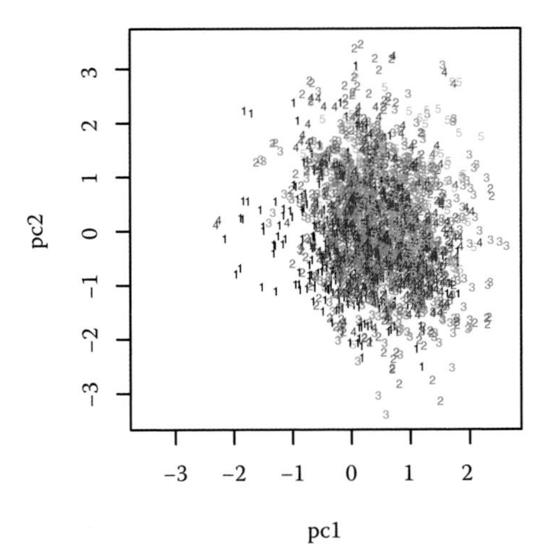

Color Figure 4.6 Plot of PC1 versus PC2 for males by age group (1: under 25; 2: 25–34; 3: 35–44; 4: 45 and over).

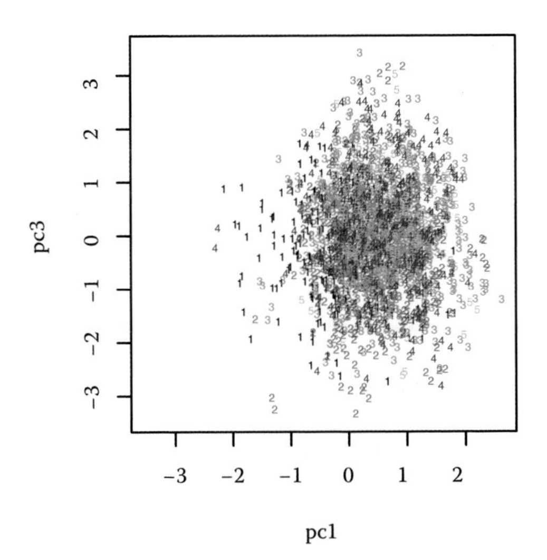

Color Figure 4.7 Plot of PC1 versus PC3 for males by age group (1: under 25; 2: 25–34; 3: 35–44; 4: 45 and over).

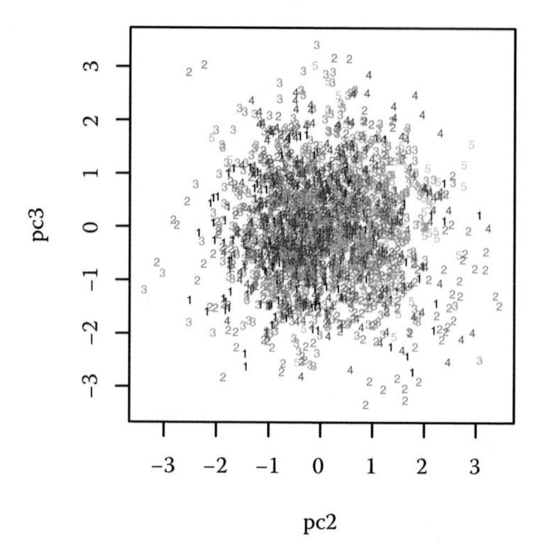

Color Figure 4.8 Plot of PC2 versus PC3 for males by age group (1: under 25; 2: 25–34; 3: 35–44; 4: 45 and over).

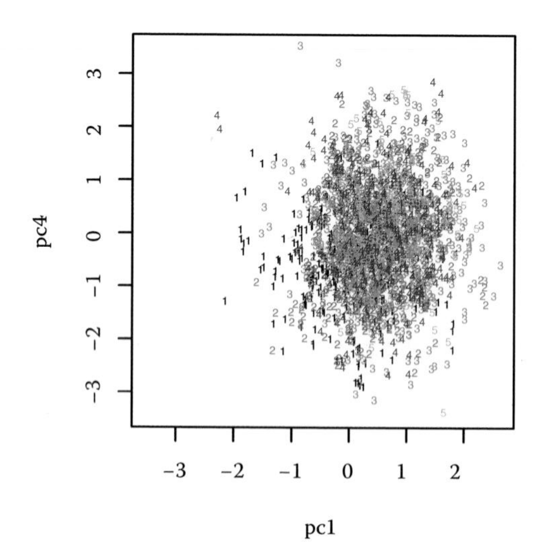

Color Figure 4.9 Plot of PC1 versus PC4 for males by age group (1: under 25; 2: 25–34; 3: 35–44; 4: 45 and over).

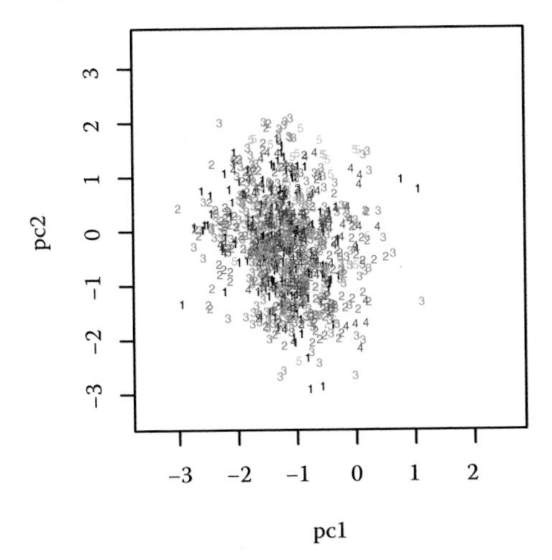

Color Figure 4.10 Plot of PC1 versus PC2 for females by age group (1: under 25; 2: 25–34; 3: 35–44; 4: 45 and over).

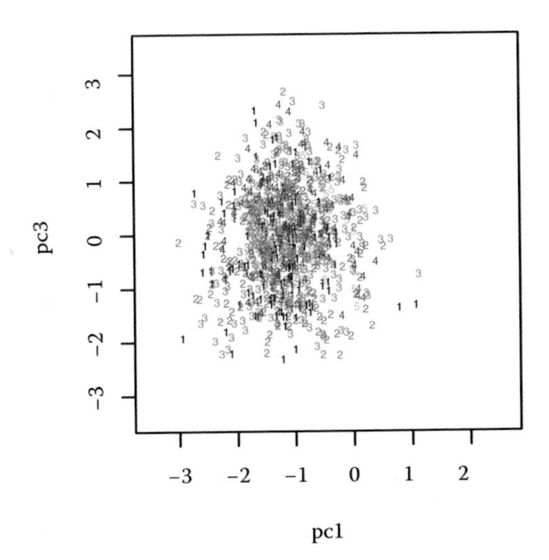

Color Figure 4.11 Plot of PC1 versus PC3 for females by age group (1: under 25; 2: 25–34; 3: 35–44; 4: 45 and over).

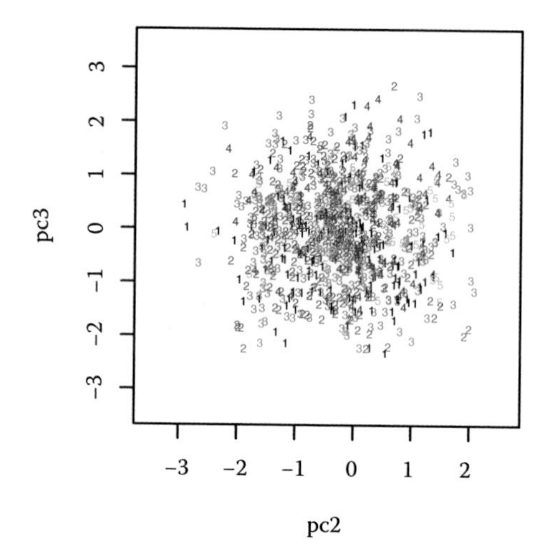

Color Figure 4.12 Plot of PC2 versus PC3 for females by age group (1: under 25; 2: 25–34; 3: 35–44; 4: 45 and over).

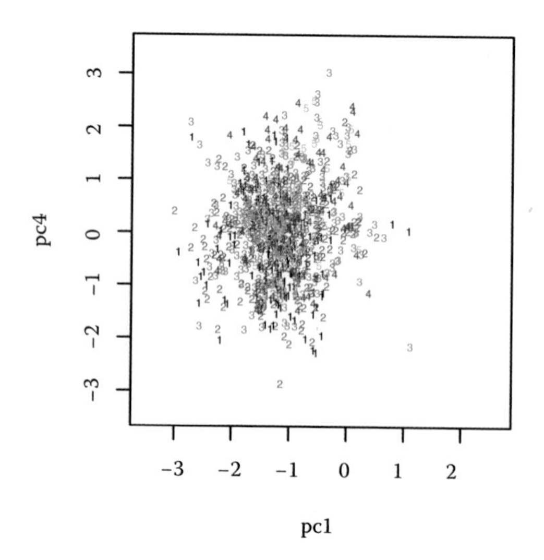

Color Figure 4.13 Plot of PC1 versus PC4 for females by age group (1: under 25; 2: 25–34; 3: 35–44; 4: 45 and over).

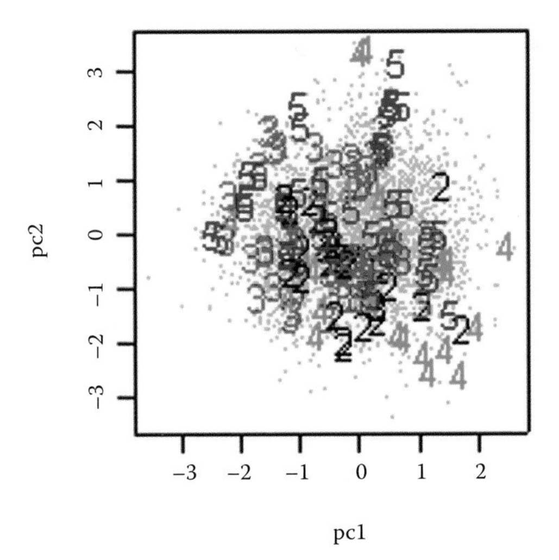

Color Figure 4.14 Plot of PC1 versus PC2 for the U.K. Census categories: White (light blue dots), Mixed (black "2"s), Asian (red "3"s), Black (green "4"s), Other (dark blue "5"s).

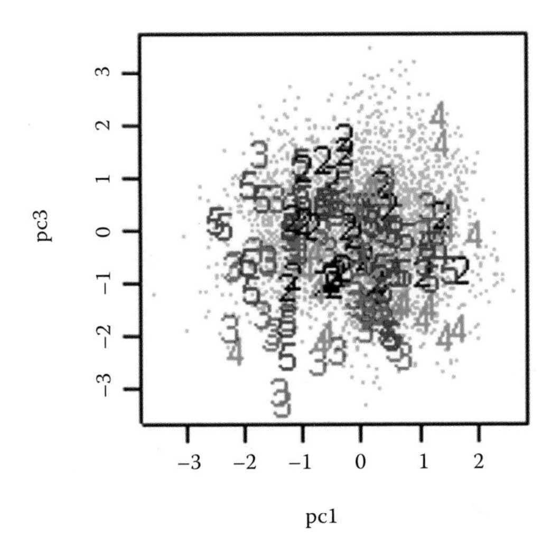

Color Figure 4.15 Plot of PC1 versus PC3 for the U.K. Census categories: White (light blue dots), Mixed (black "2"s), Asian (red "3"s), Black (green "4"s), Other (dark blue "5"s).

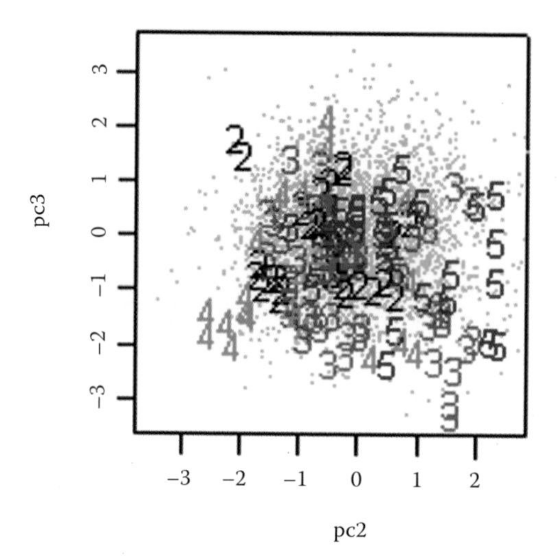

Color Figure 4.16 Plot of PC2 versus PC3 for the U.K. Census categories: White (light blue dots), Mixed (black "2"s), Asian (red "3"s), Black (green "4"s), Other (dark blue "5"s).

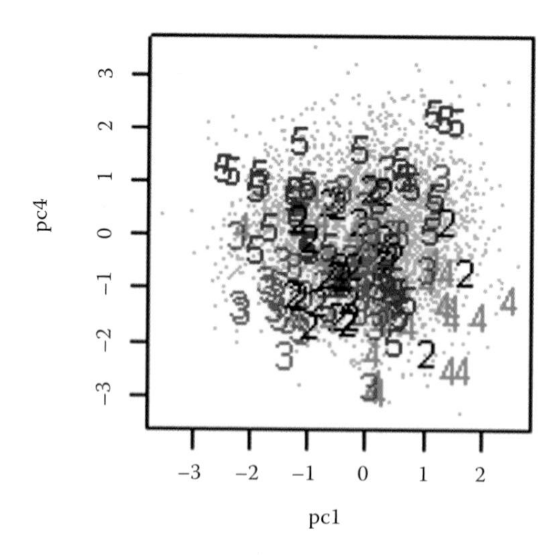

Color Figure 4.17 Plot of PC1 versus PC4 for the U.K. Census categories: White (light blue dots), Mixed (black "2"s), Asian (red "3"s), Black (green "4"s), Other (dark blue "5"s).

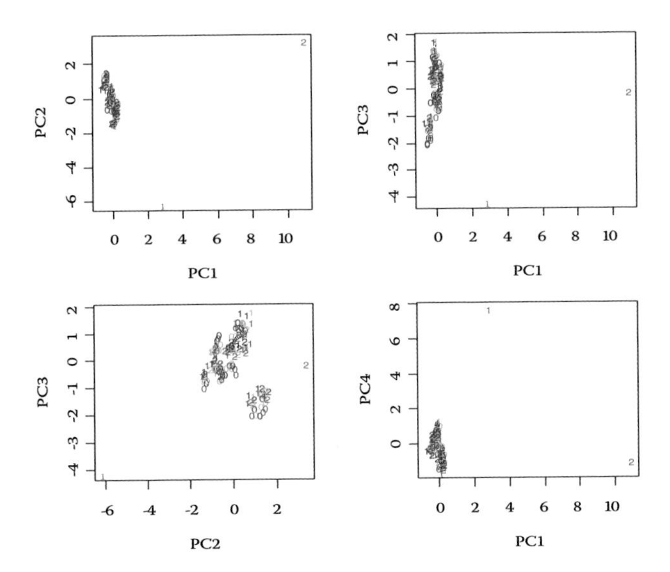

Color Figure 5.6 Results of a preliminary PCA analysis of the primary 2D facial image sample showing the first four PCs, permitting comparison of landmark datasets by subject (subjects A, B, C, D, and H) and observer (represented by three color values in the figure). Some outliers are visible as black "C"s in the figure representing observer black and subject C.

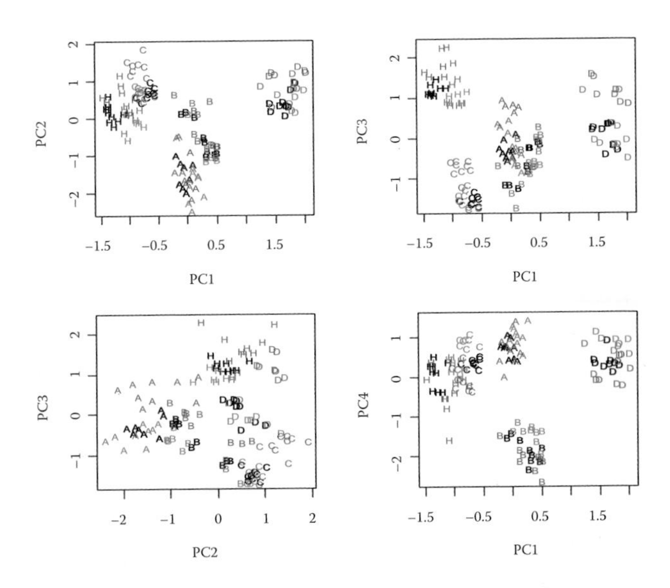

Color Figure 5.7 Results of a PCA analysis on data corrected for outliers of the primary 2D facial image sample showing the first four PCs, which account for 76.4% of variation. This analysis permits comparison of landmark datasets by subject (subjects A, B, C, D, and H) and observer (represented by three color values in the figure).

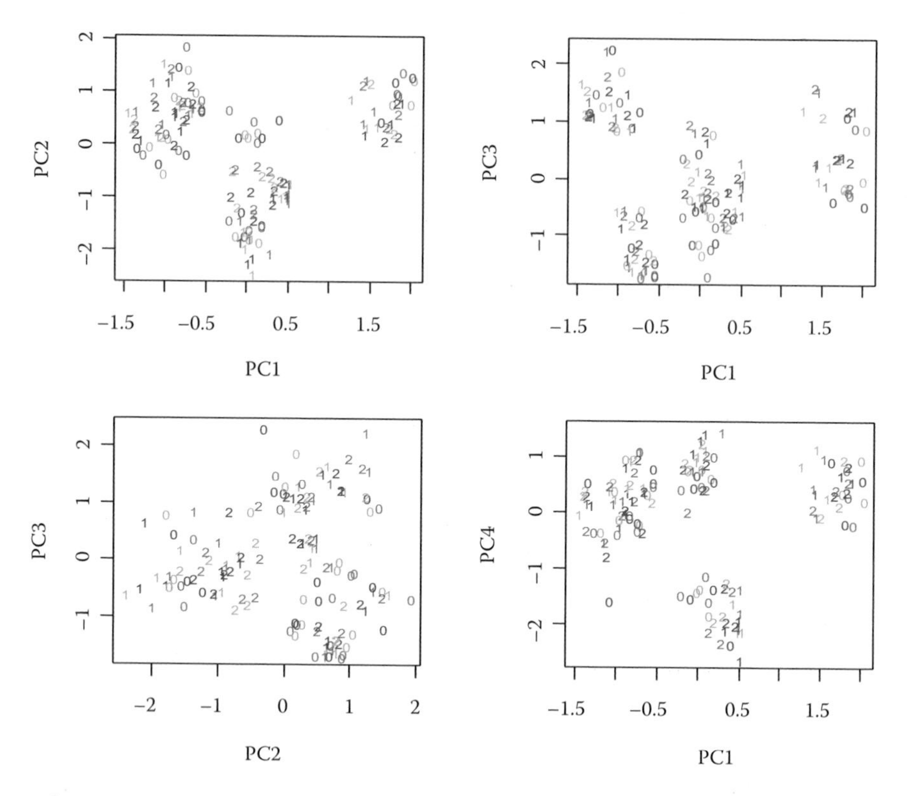

Color Figure 5.8 Results of a PCA analysis on data corrected for outliers of the primary 2D sample showing the first four PCs. This analysis permits comparison of landmark datasets by photograph of the same subject (photographs 0, 1, and 2) and landmarking repetitions (represented by three color values in the figure).

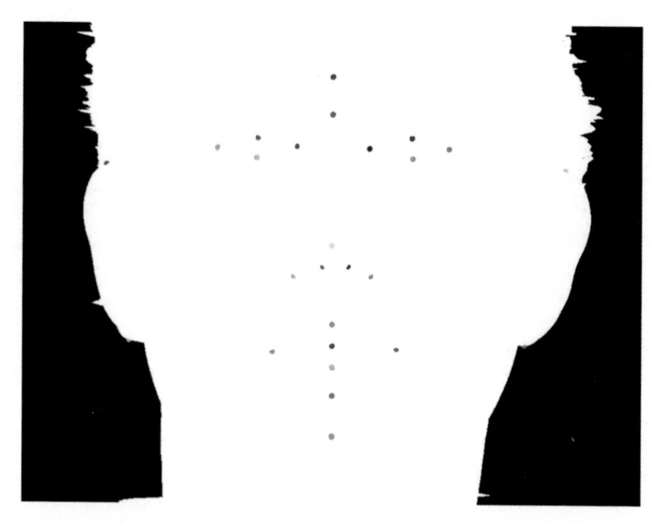

Color Figure 6.4 Frame rendered in 3ds MAX showing visible landmarks in color.

	Yaw Angle (left to right)																		
Pitch Angle	−90	−80	−70	−60	−50	−40	−30	−20	−10	0	10	20	30	40	50	60	70	80	90
−90																			
−80																			
−70																			
−60																			
−50																			
−40																			
−30																			
−20																			
−10																			
0																			
10																			
20																			
30																			
40																			
50																			
60																			
70																			
80																			
90																			

Color Figure 6.5 Landmark visibility plot for the pronasale (prn) of subject 1. The landmark is visible at all angles of pitch and yaw analyzed.

	Yaw Angle (left to right)																		
Pitch Angle	−90	−80	−70	−60	−50	−40	−30	−20	−10	0	10	20	30	40	50	60	70	80	90
−90																			
−80																			
−70																			
−60																			
−50																			
−40																			
−30																			
−20																			
−10																			
0																			
10																			
20																			
30																			
40																			
50																			
60																			
70																			
80																			
90																			

Color Figure 6.6 Landmark visibility plot for the stomion (sto) of subject 1.

	Yaw Angle (left to right)																		
Pitch Angle	−90	−80	−70	−60	−50	−40	−30	−20	−10	0	10	20	30	40	50	60	70	80	90
−90																			
−80																			
−70																			
−60																			
−50																			
−40																			
−30																			
−20																			
−10																			
0																			
10																			
20																			
30																			
40																			
50																			
60																			
70																			
80																			
90																			

Color Figure 6.7 Landmark visibility plot for the left endocanthion (en l) of subject 1. The endocanthion is visible at the smallest number angles of pitch and yaw analyzed.

	Yaw Angle (left to right)																		
Pitch Angle	−90	−80	−70	−60	−50	−40	−30	−20	−10	0	10	20	30	40	50	60	70	80	90
−90																			
−80																			
−70																			
−60																			
−50																			
−40																			
−30																			
−20																			
−10																			
0																			
10																			
20																			
30																			
40																			
50																			
60																			
70																			
80																			
90																			

Color Figure 6.8 Landmark visibility plot for the right endocanthion (en r) of subject 1. The endocanthion is visible at the smallest number angles of pitch and yaw analyzed.

	Yaw Angle (left to right)																		
Pitch Angle	−90	−80	−70	−60	−50	−40	−30	−20	−10	0	10	20	30	40	50	60	70	80	90
−90																			
−80																			
−70																			
−60																			
−50																			
−40																			
−30																			
−20																			
−10																			
0																			
10																			
20																			
30																			
40																			
50																			
60																			
70																			
80																			
90																			

Color Figure 6.9 Landmark visibility plot for the left endocanthion (en l) of subject 2. The endocanthion is visible at the smallest number angles of pitch and yaw analyzed.

	Yaw Angle (left to right)																		
Pitch Angle	−90	−80	−70	−60	−50	−40	−30	−20	−10	0	10	20	30	40	50	60	70	80	90
−90																			
−80																			
−70																			
−60																			
−50																			
−40																			
−30																			
−20																			
−10																			
0																			
10																			
20																			
30																			
40																			
50																			
60																			
70																			
80																			
90																			

Color Figure 6.10 Landmark visibility plot for the right endocanthion (en r) of subject 2. The endocanthion is visible at the smallest number angles of pitch and yaw analyzed.

	Yaw Angle (left to right)																		
Pitch Angle	−90	−80	−70	−60	−50	−40	−30	−20	−10	0	10	20	30	40	50	60	70	80	90
−90																			
−80																			
−70																			
−60																			
−50																			
−40																			
−30																			
−20																			
−10																			
0																			
10																			
20																			
30																			
40																			
50																			
60																			
70																			
80																			
90																			

Color Figure 6.11 Landmark visibility plot for the highest point of columella prime left (c′ l) of subject 1.

	Yaw Angle (left to right)																		
Pitch Angle	−90	−80	−70	−60	−50	−40	−30	−20	−10	0	10	20	30	40	50	60	70	80	90
−90																			
−80																			
−70																			
−60																			
−50																			
−40																			
−30																			
−20																			
−10																			
0																			
10																			
20																			
30																			
40																			
50																			
60																			
70																			
80																			
90																			

Color Figure 6.12 Landmark visibility plot for the highest point of columella prime right (c′ r) of subject 1.

Following end user testing, the 2D landmarking tool was used in the analysis of 2D facial images described in previous chapters by a number of operators. These investigations have yielded assessments of observer error when the tool is in use and when tried and tested it has proved robust.

The tool is a prototype anticipated to be of use in both 2D-2D and 2D-3D comparisons, and was used in combination with the application tools described below.

11.3 2D-2D Comparison Tool

This prototype addresses the requirement for a tool to compare two sets of anthropometric landmarks in 2D and produce an analysis of the likelihood of a match between the two. Behind a Microsoft Windows interface, the tool utilizes calls to modules written in R. These estimate Hotelling's two-sample T-square statistic for the two image datasets with an associated p-value, based on the analysis of the large sample database of 3D facial images collected in the study. This offers an estimation of the probability that the two landmark datasets arise from the same face. A screenshot of the 2D-2D comparison tool is shown in Figure 11.2.

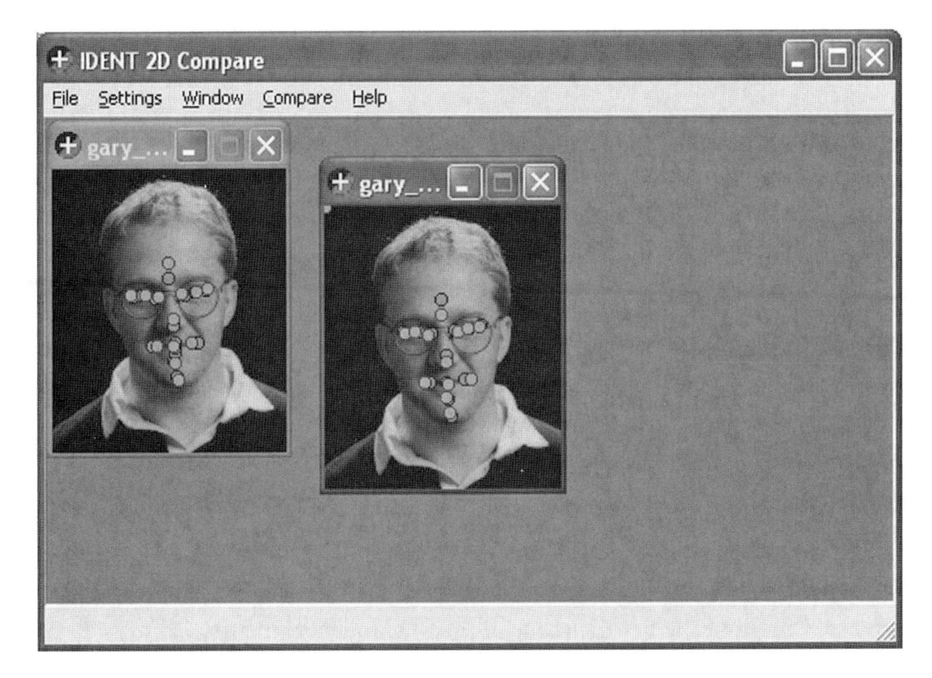

Figure 11.2 Screenshot of the 2D-2D comparison application tool.

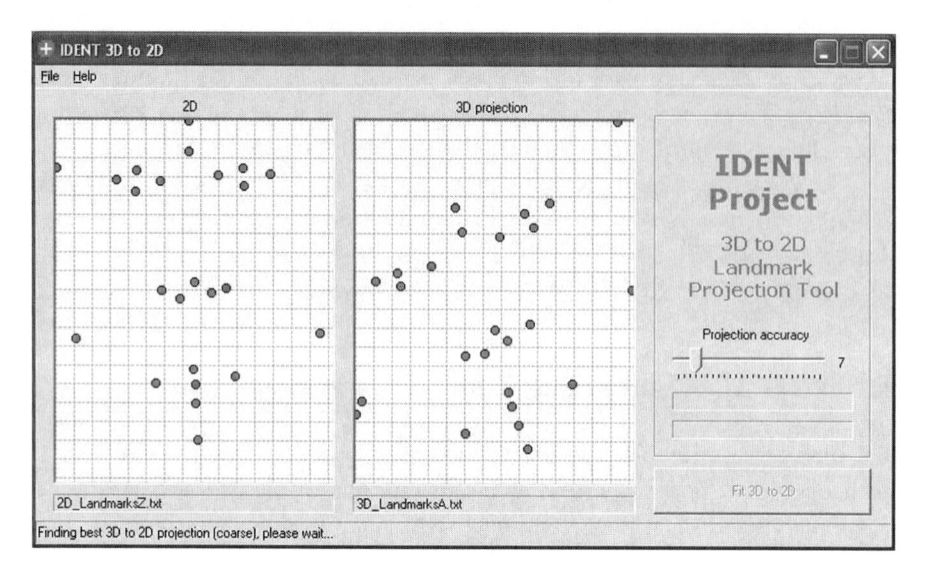

Figure 11.3 Screenshot of the 3D-2D projection tool.

11.4 3D-2D Projection Tool

In order to facilitate the scenario of face shape comparison between a 2D offender image dataset and a 3D suspect dataset—captured with the Geometrix FaceVision® FV802 System scanner, for example—an application tool was developed that would generate an optimal 2D projection of a 3D landmark dataset. A screenshot of the 3D-2D projection tool is shown in Figure 11.3.

The tool uses geometric rotation, and Procrustes transformation and scaling in R—called from Microsoft Windows application written in Visual Basic—to generate an optimal 2D projection from the 3D dataset. The optimal 2D projection dataset generated can be saved and used directly in the 2D-2D comparison tool.

Scaling makes the optimization algorithm more efficient and is valuable in case potentially matching images of the same individual have been captured at different magnifications. This is a possible disadvantage, however, as scaling could potentially lead to a false match of two different individuals whose faces are of the same shape, but different sizes!

11.5 Performance of 2D-2D and 3D-2D Comparison Systems

An assessment of the performance of the 2D landmarking tool, 2D-2D comparison tool, and 3D-2D projection tool would have been incomplete without the consideration of observer error in landmark placement in 3D and 2D, and of the effect of varying image quality. These issues have been addressed in previous chapters, and show that face shape variation in both 2D and 3D exceeds variation due to observer error under various conditions, including multiple images of the same subject. Inevitably, there is little value in making comparisons using poor images. Nevertheless, the findings consistently indicate that anthropometric landmarks can be used in 2D and 3D to reliably distinguish between subjects' facial images. Consequently, the application toolset was tested using a variety of experimental face data and face data derived from real facial comparison cases.

Testing of 2D-2D comparison tool was undertaken by means of a series of pairwise comparisons between the landmark datasets of anterior-viewed faces. Two test datasets were chosen, the first consisting of five control facial images from each of three subjects used in Chapter 5 and the second a collection of 99 random anterior photographs (not standardized control images).

An Epson Perfection® 3170 3200 dpi flatbed scanner was used to capture of analog photographic print images and transparencies. In a field-based scenario, closed-circuit television (CCTV) video surveillance images provide a prolific source of offender facial image data. An Avid Xpress® Pro (Avid, Tewksbury, MA) video editing platform with DNA Mojo uncompressed video capture card was used for the purposes of videotape and digital video image capture for the non-control faces. Ocean Systems dTective® (Ocean Systems, Burtonsville, MD) toolset was used to capture images from multiplex video and digital video formats. A Samsung® SV-5000 multisystem VCR was used for videotape input and a 42″ MX-42VM10 Maxent® plasma television for monitoring crime scene footage in analog and digital format.

All images were landmarked in 2D using the 2D landmarking tool. Each face was landmarked twice by the same operator.

Results of the first 30 comparisons attempted using the control images are given in Table 11.1. The number of face datasets used to calculate the probability of a match, taken from the large sample database, was 3272. The test used is Hotelling's T-square test, a multivariate generalization of Student's T-test. It is assumed that datasets derived from the same individual's face will share the same mean. High p-values (close to 1) indicate a high match probability and low p-values (close to 0) indicate a low match probability.

Five control images of each individual were used. This means, for example, that faces 1, 6, and 11 in Table 11.1 are of the same individual.

The results are mixed—even though the images are captured under controlled conditions of camera and camera position. Face 2 shows expected high p-values with faces 7 and 12, which are from the same subject (see Figure 11.4). Face 1, however, fails to show high p-values with faces 6 and 11—resulting in two false exclusions. False matches are also evident—such as between face 1 and faces 2, 7, 8 (see Figure 11.4) and 12, and between face 2 and faces 8 and 13.

Table 11.1 First 30 Results of Pairwise Comparisons of Control Face Data 2D-2D

Comparison	Face 1	Face 2	Number of Landmarks	Hotelling's T-Square	P-Value
1	1	2	22	0.715273	0.919549
2	1	3	22	3.35248	2.18E-12
3	1	4	23	20.4329	0
4	1	5	23	18.34483	0
5	1	6	21	9.190886	0
6	1	7	22	0.806364	0.814188
7	1	8	20	0.453457	0.998761
8	1	9	23	24.18742	0
9	1	10	21	23.74996	0
10	1	11	23	4.343488	0
11	1	12	22	0.62164	0.975835
12	1	13	22	0.243511	1
13	1	14	23	21.97175	0
14	1	15	23	20.53237	0
15	2	2	22	0	1
16	2	3	22	4.18765	0
17	2	4	22	2.34918	1.96E-06
18	2	5	22	2.671506	3E-08
19	2	6	20	1.361183	0.066627
20	2	7	22	0.520103	0.996192
21	2	8	20	0.786235	0.828888
22	2	9	22	3.246498	1.01E-11
23	2	10	20	1.319373	0.088479
24	2	11	22	0.938545	0.587798
25	2	12	22	0.184243	1
26	2	13	22	0.831225	0.776853
27	2	14	22	2.999799	3.34E-10
28	2	15	22	2.856697	2.43E-09
29	3	2	22	4.18765	0
30	3	3	24	0	1

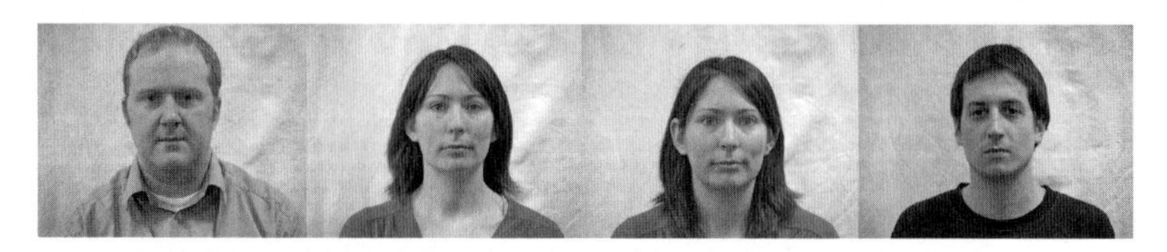

Figure 11.4 Example faces 1, 2, 7, and 8 (control data).

Results of the first 81 comparisons attempted with noncontrol facial images are given in Table 11.2. The number of face datasets used to calculate the probability of a match, taken from the large sample database, was 3258. Again, high p-values indicate a high match probability and low p-values indicate a low match probability.

The results in Table 11.2 are again mixed. Faces 1 and 2 cannot be shown for reasons of *sub judice*. Faces 3 to 17 are different images of the same individual—the male shown in Figure 11.5. The results indicate various false matches between face 1 and the individual male depicted in faces 3 to 17. The comparison of faces 1 and 2 yields a relatively low p-value indicative of a likely exclusion.

The remaining comparisons involving face 1 are with a range of noncontrol images of males and females of different ancestry, captured in different pose angles and lighting conditions, wearing different facial expressions, and with and without spectacles. Many of the remaining results in Table 11.2 yield high p-values, therefore indicative of false matches.

Testing of the 3D-2D comparison tool was pursued using a similar experimental approach for both control and random noncontrol data. For the control data, each 2D face was landmarked three times by three different operators and each 3D face was landmarked twice—once by each of two different operators. For the live data, each 2D face was landmarked twice by a single operator and each 3D face was landmarked twice—once by each of two different operators. Duplicate landmark datasets were applied in the comparison. The 2D-3D projection tool was used to project the optimal 2D projection of the 3D landmark dataset and then the 2D-2D comparison was made. The number of face datasets used to calculate the probability of a match was between 3258 and 3265.

Results of the first 16 comparisons attempted using the control dataset are given in Table 11.3. In the control dataset, the 3D face dataset originates from the male subject shown in 2D in Figure 11.4 (face 1). The 2D face datasets are from three repeat landmarkings of images from five individuals. Faces 1, 6, and 11 are from the same male individual, the others are not. The results are disappointing. None of these comparisons match this individual, apart from the final self-self comparison (face 16 with face 16); the three positive controls have generated false exclusions.

The first 28 comparisons attempted using the random noncontrol data are given in Table 11.4.

In the live dataset, all 2D face datasets are from images originating from the same individual—the male in Figure 11.5 (faces 3 and 11), and should therefore yield frequent matches with the 2D projection derived form the 3D dataset of that individual! Approximately half—13 out of 28—are false exclusions, however, only some of which can be explained by altered facial expression and so on. The explanation for these results remains to be determined and is discussed briefly below.

Table 11.2 First 81 Results of Pairwise Matching Comparisons of Noncontrol Face Data 2D-2D

Comparison	Face 1	Face 2	Number of Landmarks	Hotelling's T-Square	P-Value
1	1	2	22	1.498118	0.019615
2	1	3	20	0.748393	0.874927
3	1	4	22	0.230483	1
4	1	5	22	0.492613	0.997966
5	1	6	22	0.535126	0.994766
6	1	7	22	0.235577	1
7	1	8	22	0.571143	0.989516
8	1	9	22	0.495366	0.997827
9	1	10	22	0.518096	0.996353
10	1	11	22	0.792581	0.833392
11	1	12	22	0.678090	0.947656
12	1	13	22	0.448658	0.999353
13	1	14	22	0.746317	0.889485
14	1	15	22	0.441767	0.999469
15	1	16	22	0.535298	0.994748
16	1	17	22	0.327921	0.999992
17	1	18	22	0.733319	0.902811
18	1	19	22	0.600427	0.982669
19	1	20	22	0.178576	1
20	1	21	22	0.450256	0.999323
21	1	22	22	0.329479	0.999991
22	1	23	22	0.349456	0.999979
23	1	24	22	0.531467	0.995148
24	1	25	22	0.499779	0.997589
25	1	26	22	2.248930	6.81E-06
26	1	27	21	0.238132	1
27	1	28	13	2.402529	9.75E-05
28	1	29	22	1.404428	0.042217
29	1	30	22	0.211356	1
30	1	31	22	0.420518	0.99972
31	1	32	21	0.603640	0.979356
32	1	33	22	1.474511	0.023941
33	1	34	22	1.344494	0.066496
34	1	35	22	0.745935	0.889891
35	1	36	22	0.740753	0.895322
36	1	37	22	0.402252	0.999846
37	1	38	22	0.446514	0.999391
38	1	39	22	0.460173	0.99911
39	1	40	22	0.819399	0.795002
40	1	41	21	0.345981	0.999972

(*continued*)

Table 11.2 First 81 Results of Pairwise Matching Comparisons of Noncontrol Face Data 2D-2D (Continued)

Comparison	Face 1	Face 2	Number of Landmarks	Hotelling's T-Square	P-Value
41	1	42	22	1.417238	0.038167
42	1	43	22	1.417879	0.037974
43	1	44	21	1.606146	0.008618
44	1	45	22	1.904081	0.000369
45	1	46	22	2.534390	1.85E-07
46	1	47	22	2.381708	1.31E-06
47	1	48	22	1.921445	0.000306
48	1	49	22	5.983069	0
49	1	50	22	4.288331	0
50	1	51	22	0.696815	0.934531
51	1	52	22	0.468408	0.998892
52	1	53	22	0.872601	0.708185
53	1	54	22	0.867622	0.716814
54	1	55	20	0.375472	0.999876
55	1	56	21	8.361467	0
56	1	57	19	0.772377	0.839654
57	1	58	22	0.816701	0.799046
58	1	59	22	0.803757	0.817888
59	1	60	22	0.339790	0.999986
60	1	61	22	0.741964	0.894067
61	1	62	22	0.610929	0.979501
62	1	63	22	1.118960	0.275299
63	1	64	21	1.440091	0.034593
64	1	65	22	0.629305	0.972901
65	1	66	22	0.761487	0.872569
66	1	67	22	0.461053	0.999089
67	1	68	20	3.331185	2.29E-11
68	1	70	21	0.442996	0.999279
69	1	72	19	0.907068	0.632771
70	1	73	11	0.916647	0.572708
71	1	74	22	0.359566	0.999968
72	1	75	22	1.647872	0.005069
73	1	76	22	0.826873	0.783603
74	1	77	11	1.207665	0.230229
75	1	78	13	5.638495	0
76	1	80	20	0.370588	0.999896
77	1	81	22	0.271238	1
78	1	82	22	0.584891	0.986639
79	1	83	22	0.668695	0.953485
80	1	84	22	2.482922	3.60E-07
81	1	85	22	0.914094	0.633424

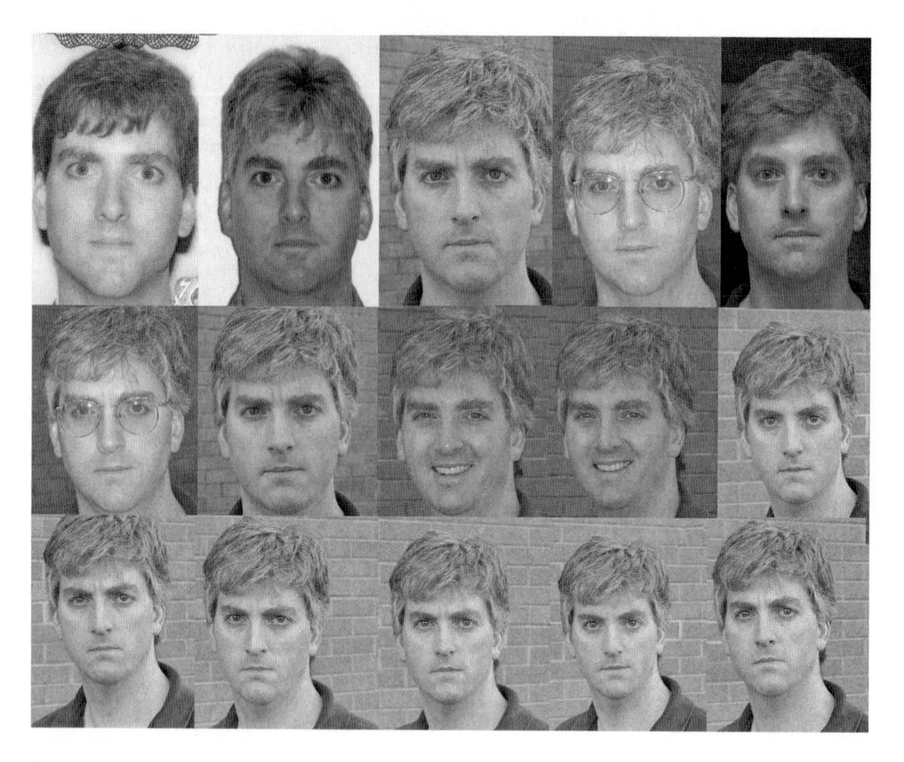

Figure 11.5 Example faces 3 to 17 (random noncontrol images).

Table 11.3 Results of First 16 Pairwise Comparisons of Control Face Data 3D-2D

Comparison	2D Face	3D Face	Number of Landmarks	Hotelling's T-Square	P-Value
1	1	16	21	14.56117	0
2	2	16	20	1.068473	0.35679
3	3	16	20	2.054334	0.000131
4	4	16	21	23.22491	0
5	5	16	21	20.02594	0
6	6	16	21	17.56574	0
7	7	16	20	1.190288	0.194385
8	8	16	20	1.158645	0.230636
9	9	16	21	25.57781	0
10	10	16	21	23.24047	0
11	11	16	21	16.6149	0
12	12	16	20	1.12907	0.268381
13	13	16	20	25.19074	0
14	14	16	21	25.19074	0
15	15	16	21	22.4378	0
16	16	16	21	0	1

Table 11.4 Results of First 28 Pairwise Comparisons of Noncontrol Face Data 3D-2D. (Face 1 is 3D and Face 2 is Uncontrolled.)

Comparison	Face 1	Face 2	Number of Landmarks	Hotelling's T-Square	P-Value
1	1	1	20	0	1
2	1	2	20	0.140266	1
3	1	3	20	0.068026	1
4	1	4	20	0.375377	0.999877
5	1	5	20	0.19401	1
6	1	6	20	0.643925	0.959272
7	1	7	20	0.255906	0.999999
8	1	8	20	0.406026	0.999668
9	2	1	20	0.140266	1
10	2	2	26	0	1
11	2	3	26	13.14468	0
12	2	4	26	11.91364	0
13	2	5	20	0.360593	0.999927
14	2	6	26	3.279179	1.34E-13
15	2	7	26	10.53013	0
16	2	8	26	10.25671	0
17	3	1	20	0.068026	1
18	3	2	26	13.14468	0
19	3	3	26	0	1
20	3	4	26	9.821604	0
21	3	5	20	0.168795	1
22	3	6	26	13.29463	0
23	3	7	26	3.368178	3.01E-14
24	3	8	26	9.761143	0
25	4	1	20	0.375377	0.999877
26	4	2	26	11.91364	0
27	4	3	26	9.821604	0
28	4	4	26	0	0

11.6 Summary

The 2D landmarking tool, 2D-2D comparison tool and 3D-2D projection tool have been developed through an iterative process of end user testing and revision, and are simple and robust user interfaces. The repeatability of the 2D landmarking tool and 3D landmark placement in Geometrix ForensicAnalyzer® has been assessed in Chapters 3 and 5, respectively. In both cases, there is consistent evidence that an anthropometric approach to distinguishing between individual subjects has promise. In this chapter, controlled assessment of performance of prototypes for 2D-2D and 3D-2D comparison was assessed.

International guidelines produced by the Biometrics Working Group (Best Practices in Testing and Reporting Performance of Biometric Devices, Version 1.0, 12 January 2000), indicate that inclusion of control and live test dataset examples of genuine and known imposter transactions should be used to evaluate the extent of false match and false nonmatch transactions.

In this case, an unacceptable level of false match and false nonmatch cases was encountered. Many, but not all, false nonmatch results may be explicable by differences in facial expression, pose, lighting, and presence or absence of glasses. Some false matches may be due to similarity in face shape between individuals. Testing was crude, however. At a processing time of approximately five minutes per 2D-2D comparison, it was concluded that a detailed calculation of false match and false nonmatch rates was unwarranted, as the levels initially encountered were high. Processing time to complete all possible 2D-2D control comparisons was estimated at 1 hour and 45 minutes. To complete all possible comparisons of random non-control data would have taken well over 80 hours.

While thorough testing remains incomplete, the prototypic tools and face databases used with them offer a user-friendly harness and valuable test data for further iterations of design, development, and testing.

Reference

Farkas, L. G. 1994. *Anthropometry of the head and face,* 2nd ed. New York: Raven Press.

Problems and Prospects 12

MARTIN PAUL EVISON AND
RICHARD W. VORDER BRUEGGE

Contents

12.1 Accomplishments and Limitations of the Research

The research and development described in this volume was ambitious in scope and completed within a demanding 24-month time frame. Collection of the large database sample was completed in 18 months.

The research encompassed scanner evaluation; evaluation of anthropometric landmarks for facial comparison in 2D and 3D; database collection and analysis, landmark visibility analysis, analysis of influence of image quality on landmark repeatability in 2D, camera error analysis, and investigation of landmark automation; prediction of missing values in the data, design and development of application tools for computer-assisted facial comparison; and investigation of courtroom admissibility issues.

The collection of a database of over 3000 facial images—each landmarked in duplicate at up to 30 sites—with consent from volunteers for confidential future research in crime prevention and detection offers an invaluable resource. Scrupulous attention was paid to ethical and consent issues in research on human subjects. The investigation of shape variation and repeatability in 2D and 3D landmarks of the face provides consistent evidence that landmarks can be used to distinguish between subjects faces.

Other findings are preliminary, but offer the following functional models:

- Model for analysis of 3D scanner image quality and accuracy
- Model for analysis of power of 2D and 3D landmarks to distinguish between faces
- Model for analysis of repeatability in computerized 2D and 3D landmark placement
- Model for analysis of effect of image quality on computerized anthropometric landmarking
- Visualization model for analysis of landmark visibility
- Visualization model for analysis of lens distortion and perspective error on landmark placement
- Model for approximation of missing data
- Model for automatic estimation of landmark position
- Prototypic application tools for 2D landmarking, 2D-2D comparison, and 3D-2D projection

These models offer much potential for further investigation.

12.2 Further Uses of the Models

12.2.1 Evaluation of Novel 3D Camera and Scanner Equipment or Landmarking Software

Biometric cameras and laser scanners are subject to constant improvement, driven by their potential uses in security and access control, television and film media, and medical imaging; the Internet is a good a starting point as any for a survey of the current 3D imaging technology.

Instruments vary in technology, size, portability, lighting method, and application software. Given the variation in ear shape and potential susceptibility to error identified in this investigation, cameras and scanners intended for computer-assisted facial comparison must be capable of accurate measurement of the ear landmarks—as well as the other landmarks. While the Geometrix FaceVision® FV802 Series Biometric Camera evaluated in this volume has been discontinued, both 3dMD™ and Cyberware® have continued to develop their ranges.

Anthropometry has a variety of application areas outside of forensic science—in medicine, human factors research and the media-visualization industries, for example—and these offer further relevant sources of information (see, for example, Enciso et al. 2004 and Gwilliam et al. 2006).

The Geometrix FaceVision FV802 Series Biometric Camera relied on ForensicAnalyzer® software to place 3D landmarks precisely by triangulation from 2D images. This approach is relatively robust compared with algorithms that synthesis the entire facial surface by matching areas of similar pixel values from different camera views, and there is no reason why future systems should not utilize standard consumer digital cameras. The cost of computing power and resolution—in megapixels—of digital cameras falls continuously.

12.2.2 Evaluation of the Influence of Image Quality and Factors Affecting Facial Appearance

The influence of image quality on 2D landmark placement was complex and further multifactorial investigation is warranted. The combination of 2D and 3D images and associated landmarks in the research databases offers a clear opportunity for the detailed investigation of the influence of pose angle on the power of landmarks to distinguish between faces. Other important factors such as facial expression, aging, and body mass index are also amenable to investigation using these methods.

When combined with the visualization models for lens distortion and perspective error, and landmark visibility, a comparative "reverse engineering" method for optimizing the position of a camera or cameras to enable the best combination of landmark visibility and distinguishing power is offered.

The disappointing outcome—given the findings consistently showing that landmarks can distinguish between subject's faces in 2D and 3D—is the high number of false match and false nonmatch results encountered when testing the prototype 2D-2D comparison and 3D-2D projection tools.

12.3 Discrimination Between Facial Images of Different Subjects

Testing of the prototypic 2D-2D comparison and 3D-2D projection tools was preliminary, and closer scrutiny of the test dataset may offer guidance to the sources of many false match and false nonmatch results. Initial results indicate that factors such as pose angle, lighting, and the presence or absence of spectacles or facial and head hair may have incurred a number of false nonmatches. Likewise, additional factors that are likely to affect any field use of the techniques developed herein include the effects of facial expression, aging, overall health, and any environmental impacts. Any or all of these

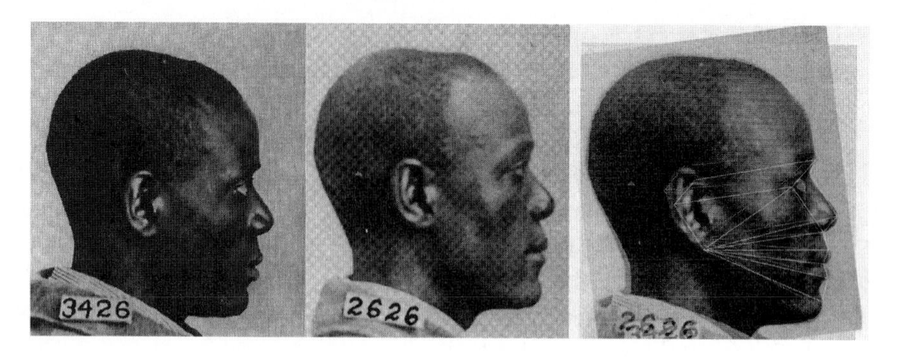

Figure 12.1 Lateral profiles and anthropometric comparison of Will and William West.

factors need to be considered. Further analysis of landmark repeatability in the test samples will also be required.

12.3.1 A Bertillon Problem?

Earlier feasibility studies (Evison 2000, Morecroft 2002) and the work of other researchers in anthropometric facial comparison (Yoshino et al. 1996, 2000, 2001; Kubota et al. 1997; Catterick 1992) has indicated that landmarks offer a promising route to distinguishing between subject's faces. For example, Evison (2000) showed that the classic profiles of Will and William West can be distinguished anthropometrically (Figure 12.1).

Morecroft (2002) showed that a small set of anterior profiles collected from the FBI Facial Identification Catalogue can be classified using six anterior 2D landmarks (Figure 12.2).

In the preliminary investigation of 61 landmarks in this study (see Chapter 4), it was established that good classification of 35 subjects landmarked three times each by two different observers could be achieved with as few as four landmarks (see Figure 12.3).

Further testing will be necessary to establish whether false positives are the simple result of a close correspondence in shape between the 2D or 2D projected from 3D face landmark datasets of different subjects. The comparison process used in this investigation is based on Procrustes registration in which landmark datasets are aligned mathematically in order to remove differences due to rotation, transposition, and scale. The initial PC analysis of the large database sample confirms that size is a major factor underlying overall variation (Chapter 4). Removing the effects of scale in the 2D-2D comparison will inevitably reduce the potential to distinguish between faces.

This introduces us to the problem encountered by the identification system of Alphonse Bertillon (1853–1914), which used 11 measurements of the

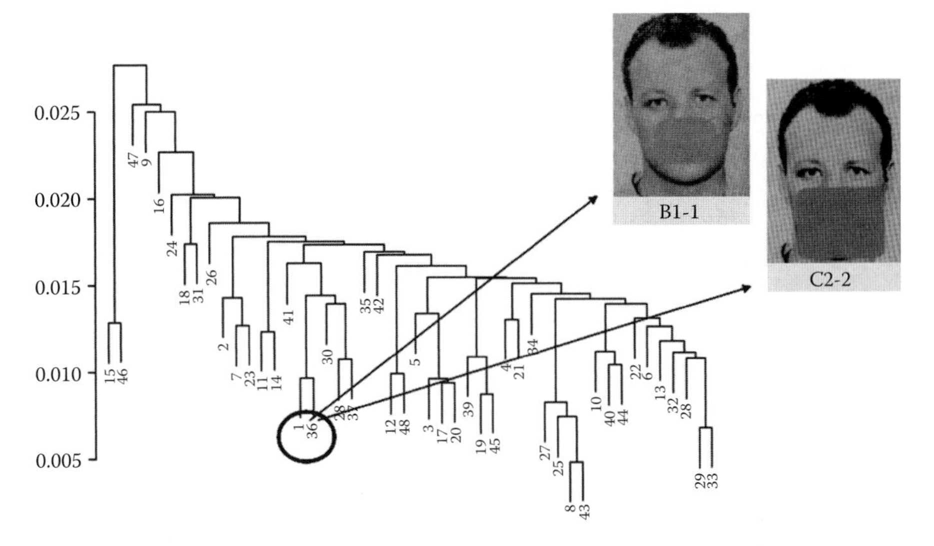

Figure 12.2 Dendrogram illustrating classification of anterior profiles from the FBI Facial Identification Catalogue using six 2D landmarks. (Figure provided by Lucy Morecroft 2002 used with permission.)

body in order to distinguish between individuals. Although Bertillonage was shown to distinguish perfectly between 241 individuals in trials, and was supported by an ingenious recording and retrieval system, both false matches and false nonmatches were encountered in practice. One apparent false match included the confusion of Will and William West—two individuals who were matched by the Bertillon method, but were distinguished by dermatoglyphic fingerprints. While factors such as measurement error and changes in dimension due to aging were shown to have the potential to underlie false nonmatches, it was the lack of independence of the variables—body measurements—that explained false matches. Many of the measurements correlated—especially due to an underlying size factor—and there was insufficient variation in the population for Bertillon's method to be effective as a means of identification. PC analysis (Chapter 4) showed factors other than size also underlie variation in the large database sample and crude estimation of frequencies (Chapter 4) indicated that probabilities would be dense in some parts of the distribution—common faces are common. Further investigation of these factors and of the independence of variables may lead to improved performance in the comparison tools.

12.3.2 Consistency with Biometrics Research

The National Academy of Sciences report (NAS 2009) mentioned in Chapter 1 not only called for the forensic sciences to be based upon a sound statistical foundation, but also noted that systematic research should be done to

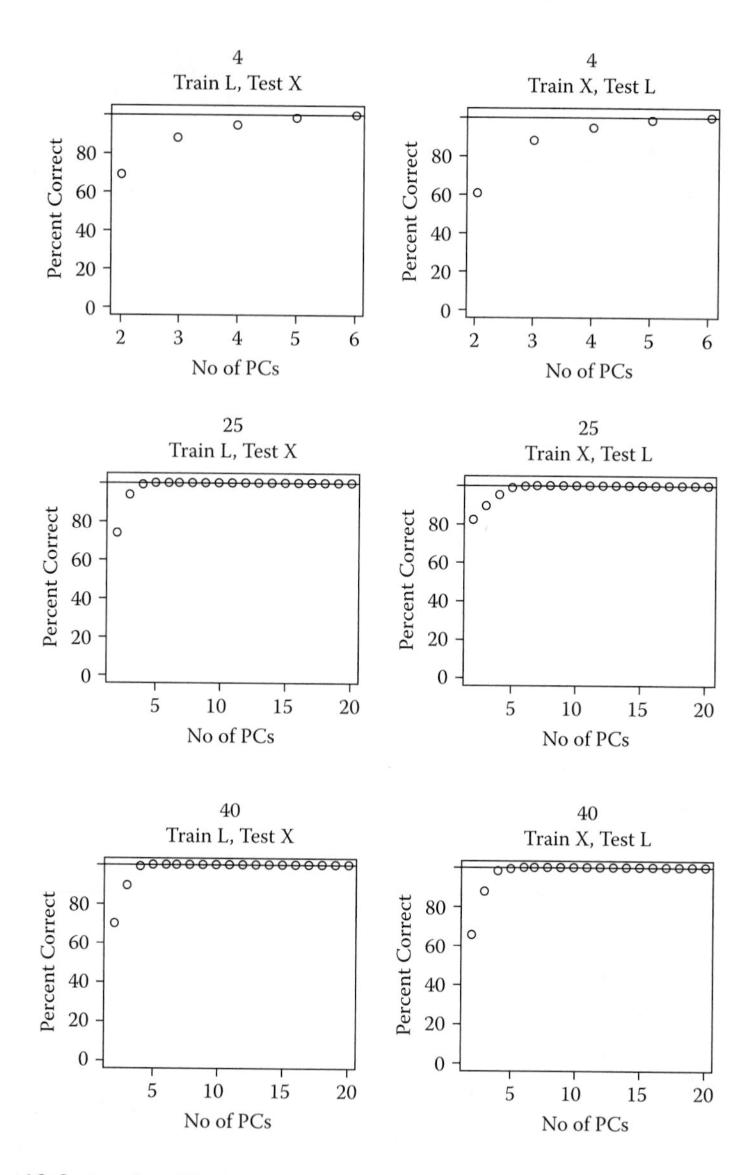

Figure 12.3 Results of discriminant analysis on up to twenty PC scores for subsets of landmarks using a training subset from one observer (L or X) to test the other. As few as four landmarks are sufficient to achieve perfect classification. Good classification is achieved with about 25, but performance begins to deteriorate with 40 landmarks or over. (Figure provided by Ian Dryden used with permission.)

validate the basic premises and techniques used in a number of disciplines. While the research described herein is unique in many ways, there are parallels to some of our findings that may be found in biometrics research—specifically in the area of facial recognition (FR)—which further validate and inform this work.

The performance of FR systems have improved by two orders of magnitude over the last 15 years (Philips et al. in press), but the forensic utility of these systems—for courtroom use—is somewhat restricted for a number of reasons. The most accurate FR results only apply to very high resolution images acquired under controlled circumstances—effectively laboratory conditions. These conditions are rarely encountered in forensic applications—whether for courtroom use or otherwise.

A further barrier to the introduction of FR system output in court is that most FR systems do not currently depend upon a physically based and validated statistical model of human faces, but instead rely upon computer learning from training sets. In other words, sets of matched and nonmatch pairs are presented to the system, and the parameters of the algorithm are adjusted to maximize the number of correct decisions and minimize the number of incorrect decisions. Such training sets are usually restricted in sample size and population distribution. Therefore, the significance of a match achieved in a given case will be limited to the size and breadth of the training set used by the system algorithm.

A further disadvantage of FR relative to anthropometric facial comparison for courtroom use is that with an anthropometric approach the visibility and position of anthropometric landmarks is clearly open to scrutiny by other experts and the court, whereas FR systems often utilize proprietary algorithms that are not immediately subject to similar scrutiny.

This is not to say that FR systems do not have a forensic utility. FR systems have already been used in numerous instances to identify wanted persons and other criminals (Mulick 2008, Indiana News 2009). But while they may help identify potential candidate matches, they are restricted because they do not provide any formal measure of the physical match between two compared subjects, nor of the frequency of a face in a representative and real world population.

The work described herein was intended to address that issue, and it does. A fundamental finding of this work is that the 3D distribution of the anthropometric landmarks studied herein, when taken alone, is unlikely to be sufficient to allow for identification of individuals to the exclusion of all others—at least when the 3D distribution is restricted to data collected during a single scan or enrollment. This finding is supported by the biometrics literature, particularly that of Bowyer et al. (2006) and Faltemier et al. (2008). Bowyer et al. summarizes multiple projects involving 3D collections and analyses, none of which provide perfect matching in 3D, while Faltemier et al. documents the improvement of 3D face recognition when multiple enrollments in 3D are incorporated.

Such work demonstrates that the 3D distribution of facial landmarks is not fixed, but can vary over time, and in such a way that the range of values displayed by a single individual will frequently overlap with those of others. Not only does the face change on a long-term scale with age, weight, and health, but facial expressions change the 3D shape of the face constantly, and this effects the 2D distribution of those features. The majority of the photographs of the subject depicted in Figure 11.5 were taken within a span of a few minutes, yet the 2D distribution of the anthropometric landmarks varied across these images. Likewise, the biometrics literature demonstrates that the 3D shape varies from day to day even when some control has been placed on expression over the course of multiple enrollments (Faltemier et al. 2008). A critical question for future research is to determine the degree to which the distribution of some or all of these 3D features can be defined within a specific range.

Even with such a determination, in order to achieve the most accurate identification results it is likely that more than a 3D shape alone will be required, if 3D is used at all. The combination of texture information in the form of a 2D image with shape information (3D) has been shown to be more effective than either technique alone (Bowyer et al. 2006). It has also been shown that recognition rates improve when multiple 2D enrollment images are used, regardless of whether they are compared to 3D systems based on a single 3D enrollment alone, or 3D combined with a single 2D enrollment (Bowyer et al. 2006). Bowyer and colleagues (2006) also indicate that approaches based on 2D images are more robust to variations in expression than 3D approaches. All of these findings support the hypothesis that skin texture is likely to be more discriminative than shape alone.

The explanation for why skin texture is so useful has not been fully investigated. Therefore, further research into the forensic utility of skin texture, including the distribution of blemishes such as freckles, moles, and scars, seems a likely area of fruitful research. Recent work by Jain and Park (2009) clearly demonstrates the utility of facial blemishes such as scars, marks, and freckles in improving the performance capability of facial recognition systems. Thus, the general ability to utilize scars, moles, tattoos, and other distinguishing features in a comparison environment is likely to reduce the frequency of false match and false nonmatch results. Much more research into this area is needed.

12.4 Admissibility Issues

Courtroom admissibility of forensic evidence (Chapter 10) remains—in general—something of a movable feast, varying from jurisdiction to jurisdiction. There are three areas of particular significance to this investigation.

12.4.1 Acceptability of the Method and the Expert

Expert evidence should be based on accepted scientific methods open to scrutiny by other experts. Novel approaches should be amenable to validation and the estimation of associated error rates. To address these criteria, the following approaches were adopted:

- Standard procedure for 2D landmarking, with error estimates
- Standard procedure for 3D landmarking, with error estimates
- Standard procedures for assessing trainee and operator performance
- Application tools that use accepted methods of multivariate statistics for comparing 2D and 3D datasets, from which false match and false nonmatch error rates can be derived
- A large database sample from which some statistical estimates of face shape frequency can be derived

While many findings are preliminary and the tools prototypic, the investigation has led to a structure within which further research and development can arise with the courtroom admissibility criteria at the fore.

12.4.2 Continuing Research and Development

The court requires that the science should be the subject of continuing research that contributes to the advancement of the technique and encompasses new technology. In addition to the further uses of the models described above, some of the methods employed—such as the R environment, active shape model, and EM algorithm—may be valuable in research at a theoretical, if not applied level.

The R Project (R 2009; see Dalgaard 2002 and Baclawski 2008 for introductions) offers an open source environment for statistical computing and graphics, which is continually enhanced by its user community. It has considerable potential for use in research and development in the multivariate analysis of the face, although other tools would have to be employed for commercial applications.

The active shape model and its cognate—the active appearance model—continue to be interesting research topics in computational biometrics (see Wen and Huang 2004, Cristinacce and Cootes 2006). Although facial recognition research takes place outside the forensic conceptual framework, it may nevertheless offer valuable insight. PCA- or EM-based methods of recreating missing data may be useful in research, but they are unlikely to be acceptable in court: a statistical estimate of the probable position of an anthropometric landmark on the face seems as unattractive a courtroom prospect as one of the location and type of a fingerprint minutia, or of the alleles missing from a partial DNA profile.

A semantic or rule-based approach to automated landmark placement would be more appropriate for the courtroom and, given the unambiguous and unique 3D topography of the human face, could yield initial rapid benefits. While reducing the opportunity for human error, a rule-based system derived from 3D facial anatomy would be far more transparent to the court than a method based on multivariate statistics—however elegant that would undoubtedly be.

Likewise, given the likely importance of texture information in the human identification from images problem, reliable approaches to locate, extract, and analyze the distribution of 2D skin texture features are sorely needed. Automated methods for doing this would make it easier to conduct studies of large populations, which would then permit the development of statistical models to provide an objective measure of a match probability.

The visualization models are not only valuable research tools in their own right, they offer a route to the courtroom visualization of evidence in facial comparison, which may be invaluable in communicating analysis and findings to the jury (see March et al. 2004 for an example).

12.4.3 The Concept of Human Identification

Much of forensic science—it can be argued—has been based on a concept of discernable uniqueness that is fallacious (see Saks and Koehler 2005). The science of forensic DNA profiling is based on well understood and empirically supported models of molecular and population genetics, but the outcome of its analysis is a match probability. Although the match probability may exceed 1 in 1 billion, it is not a claim of uniqueness or identification—although a lawyer may imply it amounts to it. Despite the long and prestigious history of dermatoglyphic fingerprinting, it offers no theoretical or statistical equivalent to the match probability of DNA profiling.

This investigation has not assumed the uniqueness of the face—however intuitive it would be to make that assumption—or the uniqueness of a 3D facial anthropometric landmark dataset. It has sought to empirically examine face shape variation using an anthropometric approach, which would offer an empirical foundation for statistical estimates of face shape—or landmark dataset—frequency. Whether measurable variation will be sufficient to allow powerful odds to be offered to the court or whether the facial landmarks will—like the bodily dimensions used by Bertillon—be frequently encountered, remains an open question. Even if the latter is true, it will still be valuable to the court to know the facial equivalent of a match probability. The ultimate problem of Bertillonage was that there was a convenient and better alternative—the fingerprint. There is no such equivalent in a crime scene facial image.

Experts do make errors (see Dror and Charlton 2006 for a discussion regarding dermatoglyphic fingerprinting). A number of empirically supported methods for measurement of error have been offered in this investigation, as well as procedures that would permit training in and quality assurance of the analytical methods explored. The computational tools are a route to the reduction of human error and their further automation would enhance this.

12.5 Dissemination of the Database and Project Materials

The complete large facial image database sample is available for dissemination by intergovernmental agreement, on the proviso that the conditions under which the sample was collected are respected. These are reflected in the information sheet and consent form provided to the volunteers (see Appendix A) and include a requirement for confidentiality—the database images must not be published or otherwise disseminated—and for the images to be used for the purpose for which consent was given—namely, for research in crime prevention and detection.

12.6 Conclusion

Inevitably, research has raised as many questions as it has answered. The structure of this investigation, however, is such that much future work can be based on it or it potentially offers more promising avenues that may be explored.

References

Baclawski, K., 2008. *Introduction to probability with R.* Boca Raton, FL: Chapman & Hall.

Bowyer, K. W., K. Chang, and P. J. Flynn, 2006. A survey of approaches and challenges in 3D and multi-modal 3D+2D face recognition. *Comput. Vis. Image Understanding* 101(1): 1–15.

Catterick, T., 1992. Facial measurements as an aid to recognition. *Forensic Sci. Int.* 56: 23–27.

Cristinacce, D. and T. F. Cootes, 2006. Facial feature detection and tracking with automatic template selection. *Proc. 7th IEEE Int. Conf. on Automatic Face and Gesture Recognition* 2006: 429–34.

Dalgaard, P., 2002. *Introductory statistics with R.* New York: Springer.

Dror, I. E. and D. Charlton, 2006. Why experts make errors. *J. Forensic Identification* 56(4): 600–16.

Enciso, R., E. S. Alexandroni, K. Benyamein, R. A. Keim, and J. Mah, 2004. Precision, repeatability and validation of indirect 3D anthropometric measurements with light-based imaging techniques. *IEEE Int. Symp. on Biomedical Imaging: Nano to Macro* 2004 2: 1119–22.

Evison, M. P., 2000. Anthropometry of the face. A review of the traditional methods of craniofacial measurement and their application to anthropometry of photographic images. *Third UK National Conference on Craniofacial Identification Report on Proc.* Department of Art in Medicine, University of Manchester, p. 7 (abstr.).

Faltemier, T. C., K. W. Bowyer, and P. J. Flynn, 2008. Using multi-instance enrollment to improve performance of 3D face recognition. *Comput. Vis. Image Understanding* 112(2): 114–25.

Gwilliam, J. R., S. J. Cunningham, and T. Hutton, 2006. Reproducibility of soft tissue landmarks on three-dimensional facial scans. *Eur. J. Orthodontics* 28:408–15.

Indiana News. 2009. Facial recognition system leads to fraud arrest. *Indiana News*, June 2, 2009. http://www.theindychannel.com/news/19635290/detail.html (accessed June 30, 2009).

Jain, A. K., and U. Park, 2009. Facial marks: Soft biometric for face recognition. *MSU Technical Report*, MSU-CSE-09-10, 4 pp.

Kubota, S., H. Matsuda, K. Imaizumi, S. Miyasaka, and M. Yoshino, 1997. Anthropometric measurement and superimposition technique for facial image comparison using 3D morphologic analysis, *Rep. Natl. Res. Inst. Police Sci.* 50:88–95 (in Japanese).

March, J., D. Schofield, M. P. Evison, and N. Woodford, 2004. Three-dimensional computer visualization of forensic pathology data. *Am. J. Forensic Med. Pathol.* 25(1): 60–70.

Morecroft, L. 2002. Identification of faces. MSc diss., The University of Sheffield.

Mulick, S. 2008. Facial recognition software gives Pierce County help in tough cases. *The Tacoma News Tribune*, December 22, 2008. http://www.seattlepi.com/local/393308_computercrime23.html (accessed June 30, 2009).

NAS, 2009. *Strengthening forensic science in the United States: A path forward.* Washington DC: National Research Council, 254 pp.

Philips, P. J., W. T. Scruggs, A. O'Toole, P. J. Flynn, K. W. Bowyer, C. L. Schott, and M. Sharpe, 2009 FRVT 2006 and ICE 2006 large-scale experimental results. *IEEE Trans. Pattern Anal. Machine Intelligence,* DOI 10.1109/TPAMI. 2009.59.

R., 2009. *The R project for statistical computing.* http://www.r-project.org (accessed June 30, 2009).

Saks, M. J., and J. J. Koehler, 2005. The coming paradigm shift in forensic science. *Science* 309: 891–5.

Wen, Z., and T. S. Huang, 2004. *3D face processing.* Boston: Kluwer.

Yoshino, M., S. Kubota, H. Matsuda, K. Imaizumi, S. Miyasaka, and S. Seta, 1996. Face-to-face video superimposition using three dimensional physiognomic analysis, *Jpn. J. Sci. Technol. Identification* 1: 11–20.

Yoshino, M., H. Matsuda, S. Kubota, K. Imaizumi, and S. Miyasaka, 2000. Computer-assisted facial image identification system using 3D physiognomic range finder, *Forensic Sci. Int.* 109: 225–37.

Yoshino, M., H. Matsuda, S. Kubota, K. Imaizumi, and S. Miyasaka, 2001. Computer-assisted facial image identification system. *Forensic Sci. Commun.* 3(1). http://www.fbi.gov/h/lab/fsc/backissu/jan2001/yoshino.htm (accessed December 8, 2009).

Appendix A

Information Sheet, Biographic Form, and Consent Form

Computer-Assisted Facial Comparison Project Information Sheet

Contents

A.1 Information Sheet

Computer-Assisted Facial Comparison Project
Information Sheet

You are being asked to participate in a research study. Before you decide, it is important for you to understand why the research is being done and what it will involve. Please take time to read the following information carefully and discuss it with others if you wish. Ask us if there is anything that is not clear or if you would like more information. Take time to decide whether or not you wish to take part.

What is the purpose of this study?
The faces of perpetrators of often-serious crimes are regularly caught on Closed Circuit Television (CCTV) or security cameras. You may have seen such images on television on *BBC Crimewatch UK* or *Crime stoppers*. Little scientific work has been done that can be used to help the police or a jury decide whether the face in the CCTV image is the same as that of someone suspected of committing a crime. We think there is some danger of both unwarranted convictions and acquittals in cases involving facial comparison.

This research project will develop a means of comparing faces. This will be accomplished by measuring the distances between certain points on the face. We will be using the information from your 3D photograph to take these measurements. This will help us understand how people's faces vary, and help us to create better methods of recognizing and distinguishing between faces scientifically.

Who is organizing and funding the research?
The University of Sheffield is organizing the research project with other scientists at Nottingham and Kent Universities. An international consortium of agencies whose goal is the prevention and detection of crime, and/or the administration of justice sponsor the project—including the US Federal Bureau of Investigation. The project will follow the guidelines of the UK Police Information Technology Organisation. Authority for this research project rests with the University of Sheffield.

Why have I been chosen?
We are gathering information from volunteers who are willing to participate in our project.

Do I have to take part?
Participation in the project is purely voluntary. It is up to you whether or not you take part. If you do decide to take part you will be given this information sheet to keep and be asked to sign a consent form.

DESTRUCTION NOTICE – Destroy by any method that will prevent disclosure of contents or reconstruction of the document to any person or entity not authorized such access by the contractor or the Government

Computer-Assisted Facial Comparison Project
Information Sheet

What will happen to me if I take part?

We will take your 3D photograph and record the following biographical information: your age, sex, ancestral affiliation (ethnicity), and whether any of your relatives are also volunteering. We need to record this information as these factors can affect face shape. The 3D photograph and biographic information will be kept in a secure database. The sponsors will routinely be provided access to and keep this 3D photograph and biographical information database after the project is over so that it can continue to be used by researchers interested in crime prevention and detection. It will not be used for any purpose other than scientific and technical research. If you initially decide to take part you are still free to withdraw at any time without giving a reason and your database record will be destroyed.

In addition to the biographical information identified above, we will also record your name. We need to record your name in case you ask us to remove your data later on. If you want further information about the project we will also record your email address. Your name and email address (if you provide it) will be stored in a secure database that will be separated from the database containing your 3D photograph and biographical information. The University of Sheffield organizers will maintain control and access to this separate database. A unique key, allocated by the University of Sheffield researchers, will reside in both databases, providing us with the ability to destroy your database record, should you request it.

Except as described above, your 3D photograph will not be made public or distributed outside of the scientific, technical or research community and we will not publish any other personal information that will allow you to be specifically identified with your 3D photograph. Your name and email address will not be made public or distributed beyond the Sheffield University researchers engaged in this project.

What will happen to the results of the research study?

The results will be part of a research project due to be completed in Autumn 2005. The scientific results will be published and it is intended that new tools for comparing faces which result from this research will be made available to police, defence lawyers and courts.

Who has reviewed this study?

The Research Ethics Committee at the University of Sheffield has reviewed this study.

Contact for further information:

Dr Martin Evison, Department of Forensic Pathology, The University of Sheffield, The Medico-Legal Centre, Watery Street, Sheffield, S3 7ES, United Kingdom. Tel. +44 114 2738721, Fax. +44 114 2798942, Email. m.p.evison@sheffield.ac.uk.

DESTRUCTION NOTICE – Destroy by any method that will prevent disclosure of contents or reconstruction of the document to any person or entity not authorized such access by the contractor or the Government

A.2 Biographic Form

Computer-Assisted Facial Comparison Project
Biographic Details
CONFIDENTIAL

Please provide the following information:-

Age: ☐☐ Sex (m / f): ☐

How would you describe your ancestry / ethnicity?

Please tick ✓ in the box that applies to you.

White

British ☐ 01

Any other White background (Please describe) ☐ 02 — *Please describe*

Mixed

White and Black Caribbean ☐ 03

White and Black African ☐ 04

White and Asian ☐ 05

Any other Mixed background (Please describe) ☐ 06 — *Please describe*

Asian or Asian British

Indian ☐ 07

Pakistani ☐ 08

Bangladeshi ☐ 09

Any other Asian background (Please describe) ☐ 10 — *Please describe*

Black or Black British

Caribbean ☐ 11

African ☐ 12

Any other Black background (Please describe) ☐ 13 — *Please describe*

Chinese or other ethnic group

Chinese ☐ 14

Any other (Please describe) ☐ 15 — *Please describe*

Are any of your relatives participating in this study? Please give details:

Name	Date of Birth (d d m m y y)	They are your? (e.g. mother, brother)

DESTRUCTION NOTICE – Destroy by any method that will prevent disclosure of contents or reconstruction of the document to any person or entity not authorized such access by the contractor or the Government

A.3 Consent Form

Computer-Assisted Facial Comparison Project

Consent Form

CONFIDENTIAL

Family Name: ☐☐☐☐☐☐☐☐☐☐☐☐☐☐☐☐☐☐☐☐☐☐☐☐☐☐☐

First Name: ☐☐☐☐☐☐☐☐☐☐☐☐☐☐☐☐☐☐☐☐☐☐☐☐☐☐☐

d d m m y y

Date of Birth: ☐☐☐☐☐☐

Please read the statement and tick ✓ in the boxes that apply to you:-

I have read and understood the above information. I consent to the collection, long term storage and retention and use of my image and biographic information for scientific and technical research in the UK and elsewhere. I certify that the information disclosed by me is true and accurate to the best of my knowledge. ☐

I would like to be kept informed about this study. I agree for my email address to be kept on computer by the University of Sheffield in order to facilitate their disclosure of periodic updates, time and resources permitting. I understand that this e-mail information will be stored separately and will be controlled and accessed only by the organizers of this project, and will not be distributed to any other party. ☐

Email: ☐☐☐☐☐☐☐☐☐☐☐☐☐☐☐☐☐☐☐☐☐☐☐☐☐☐☐☐

Signature: ________________________

Signature of Parent or Guardian if under 18: ________________

Date: __________

Thank you for participating in this study!

For project use only

Scanner: ☐ 1 Geometrix 2 Cyberware 3 Both

Location: ☐ 1 Magna

Key: 0☐☐☐☐

Date: ☐☐☐☐☐☐

d d m m y y

Operator: ☐☐☐

DESTRUCTION NOTICE – Destroy by any method that will prevent disclosure of contents or reconstruction of the document to any person or entity not authorized such access by the contractor or the Government

Appendix B

Companion DVDs

Contents

B.1 Introduction

Two companion DVDs are included to foster further research in forensic facial comparison. The DVDs includes much of the raw image and landmark data used in the various chapters, as well as source code in R, MATLAB®, and MAXScript® used in the analyses. Many output data files are also included, including spreadsheets in which cumulative data is derived and presented in chart form. The source code of prototypic tools for 2D landmarking, and facial comparison from 2D-2D and 2D-3D are included. The data, code, results, and charts are offered in a far more comprehensive form than is possible in the printed version. Additional presentation material, including video clips illustrating 3D landmark visibility and other demonstrative examples are also included.

The material included in the companion DVDs is partial and prototypic, and intended for further testing, use in research, and model building only.

B.2 Contents by Folder

B.2.1 Chapter 1

This folder includes three spreadsheets containing the final raw landmark data, including a version with missing data estimated using the EM algorithm. Also included are one spreadsheet containing the biographic data corresponding to each landmark dataset, and one file listing the 30 landmarks, their numbers, and names.

B.2.2 Chapter 2

This folder contains the raw data, and scripts and spreadsheets used in the analyses of the Geometrix®, Cyberware®, and 3dMD® scanners.

B.2.3 Chapter 3

This folder contains R source code used in the analysis of landmark variation in 3D, incorporating matching of landmark dataset and biographic records, and includes instructions on calling R from Windows.

B.2.4 Chapter 4

The tools used in analysis of the large sample database were based on those employed in Chapter 3.

B.2.5 Chapter 5

This folder contains the images and the raw landmark used in the investigation of anthropometric landmarking in 2D. The 2D landmarking tool is also included.

B.2.6 Chapter 6

This folder contains the 3D face image scan data and supporting material used in the study of the effect of 3D rotation on landmark visibility, including all 90 color grids used to depict landmark visibility and AVI visualizations of 3D landmark projections in rotation in vivo, and other material.

B.2.7 Chapter 7

This folder contains the 3D landmark data and supporting material used in the study of the influence of lens distortion and perspective error. Spreadsheets, 3ds files, and MAXScript sources used in the "deep head" analysis are included.

B.2.8 Chapter 8

This folder contains MATLAB scripts and supporting data used in the estimation of landmark position using the active shape model.

B.2.9 Chapter 9

There are no tools associated with this chapter (Generation of Values for Missing Data).

B.2.10 Chapter 10

There are no tools associated with this chapter (Admissibility).

B.2.11 Chapter 11

This folder contains source code (in Visual Basic and R), binaries and compiled versions of the prototypic application tools.

B.2.12 Chapter 12

There are no tools directly associated with this chapter (Problems and Prospects); some results are presented derived from the tools used in Chapter 3.

B.2.13 Facial Expression Dataset

A "bonus" dataset is included containing images of six subjects photographed three times each wearing normal, happy, surprised, and angry facial expressions, captured in BMP, NEF, TIF, and JPG formats.

B.2.14 Project Documents

This folder contains the Information Sheet, Biographic Form, and Consent Form used in data collection in Word format (shown in Appendix A). The Landmarking Protocol used to locate the 30 anthropometric landmarks used in data collection is also included, in both PDF and MHT (HTML) formats.

Index